Cancer Preventive and Therapeutic Compounds: Gift From Mother Nature

Edited by

Sahdeo Prasad & Amit Kumar Tyagi

Department of Experimental Therapeutics, Division of Cancer Medicine, The University of Texas M. D. Anderson Cancer Center, Houston, TX 77054, USA

Cancer Preventive and Therapeutic Compounds: Gift from Mother Nature

Editors: Sahdeo Prasad and Amit Kumar Tyagi

eISBN (Online): 978-1-68108-491-6

ISBN (Print): 978-1-68108-492-3

First published in 2017.

liability of Bentham Science Publishers shall be limited to the amount actually paid by you for the Work.

General:

1. Any dispute or claim arising out of or in connection with this License Agreement or the Work (including non-contractual disputes or claims) will be governed by and construed in accordance with the laws of the U.A.E. as applied in the Emirate of Dubai. Each party agrees that the courts of the Emirate of Dubai shall have exclusive jurisdiction to settle any dispute or claim arising out of or in connection with this License Agreement or the Work (including non-contractual disputes or claims).

2. Your rights under this License Agreement will automatically terminate without notice and without the need for a court order if at any point you breach any terms of this License Agreement. In no event will any delay or failure by Bentham Science Publishers in enforcing your compliance with this License Agreement constitute a waiver of any of its rights.

3. You acknowledge that you have read this License Agreement, and agree to be bound by its terms and conditions. To the extent that any other terms and conditions presented on any website of Bentham Science Publishers conflict with, or are inconsistent with, the terms and conditions set out in this License Agreement, you acknowledge that the terms and conditions set out in this License Agreement shall prevail.

Bentham Science Publishers Ltd.
Executive Suite Y - 2
PO Box 7917, Saif Zone
Sharjah, U.A.E.
Email: subscriptions@benthamscience.org

BENTHAM SCIENCE

CONTENTS

FOREWORD

Among various chronic diseases, cancer is one of the most dreaded diseases throughout the world. Despite dramatic improvements in surgical and reconstructive techniques, the overall mortality rates for cancer remain relatively unchanged. To date, numerous screening and preventive approaches have been directed towards cancer, which clearly reflect a decrease in the morbidity and mortality associated with cancer. The screening measures include physical exam and history, laboratory tests, imaging and genetic tests on timely basis. However, primary prevention of cancer is more important to keep cancer away from developing. This includes maintaining a healthy lifestyle and avoiding exposure to known cancer-causing substances. Thus cancer risk can be reduced with healthy choices like avoiding tobacco, limiting alcohol use, protecting skin from the sun and avoiding indoor tanning, eating a diet rich in fruits and vegetables, keeping a healthy weight, and being physically active.

Lifestyle plays an important role in the prevention of this disease. By adopting a diet consisting primarily of whole grains, fruits and vegetables with limited amounts of meat, primarily chicken or cold-water fish, and doing daily exercise, several risk factors can be avoided. Various methods are available for the treatment of cancers and the selection will depend on the cost, morbidity, requirement of reliable biopsy specimens, resources available, etc. Thus management of cancer by self-care could be a great potential to improve detection and the treatment of cancer; morbidity and mortality also will decrease as a result. The field has broad and wide applications and with every new development reported in leading peer-reviewed journals across the globe, the opportunities only become wider and the hopes brighter.

The editors Dr. Sahdeo Prasad and Dr. Amit Tyagi have done an excellent job of bringing out timely peer-reviewed chapters under the banner "*Cancer Preventive and Therapeutic Compounds: Gift From Mother Nature*" with contributors spreading across four different continents. I complement the authors and appreciate their efforts in bringing out this comprehensive compilation. It is quite impressive to note that the editors have tried to capture such a wide and dynamic topic in a series of attractive articles highlighting different forms of cancer prevention and treatment research, both existing and newly emerging technologies in the field, approaches, advantages, thoughts from around the world along with potential future prospects. The simplicity of the language and presentation style is very much appealing and impressive.

It is my great pleasure to pen down/write the foreword for this prestigious, multi-authored book compilation withpeer-reviewed chapters. This book will be a valuable resource for the scientists and students seeking updated and critical information for their experimental plans. It will be very useful for the clinician to develop clinical trials using natural compounds with or without existing therapeutic drugs. Pharmaceutical companies could design new formulations based on the literature available in this new book. Most importantly, normal population and cancer patients can be benefited by knowing the preventive and therapeutic efficacy of natural compounds. They can use these natural compounds in their routine life. This book could be a major breakthrough worldwide for the readers, particularly the cancer patients.

Anushree Malik, PhD
Associate Professor
Applied Microbiology Lab, CRDT
Indian Institute of Technology Delhi, New Delhi
India 110016

PREFACE

Cancer is one of the leading causes of deaths around the world and it is globally increasing. The highest incidence rates are in the developed countries such as the USA, and the lowest rates are found in developing countries. These differences in incidence rates appear to be attributable to geographical differences in diet and environmental exposure. Although environmental and genetic factors are the major risk factors for cancer, lifestyle also contributes to the development of this disease. Although screening modalities for early detection and therapeutic management of cancer have improved considerably, this disease still needs better treatment modalities. Since long-term use of cytotoxic chemotherapy and radiotherapy can have severe side effects and since tumors can develop resistance to these therapies, agents that can overcome tumor resistance, enhance the therapeutic efficacy of existing drugs and can control multiple signaling pathways are needed to treat cancer.

Although numerous anticancer drugs are available, most of them are expensive and have serious side effects. Thus, the challenging task of finding an alternative cancer treatment measure has become more important than ever to both scientists and physicians. Since natural compounds have been identified and explored for their health benefits for centuries, several nutritional factors have attracted considerable attention as modifiable risk factors in the prevention and treatment of cancer. Natural products are important sources of anti-cancer lead molecules; even many successful anti-cancer drugs approved by FDA are natural products or their derivatives. Still many more are under clinical trials. Based on the current available research, the present book will focus on the chemopreventive and anti-cancer activities of different natural/dietary compounds such as fruits, vegetable, spices, legumes, nuts, grains, and cereals highlighting their potential use against cancer treatment. Since these natural compounds including fruits and vegetables contain a wide variety of phytochemicals, they may have anti-carcinogenic effects. Evidences showed that the phytochemicals present in fruits and vegetables modulate large numbers of cell signaling molecules linked with cancer. The modulation of signaling molecules controls the abnormal growth of cells and ultimately controls the growth of cancer. Also antioxidative and anti-inflammatory properties of natural compounds could hold promise for cancer chemoprevention because oxidative stress and chronic inflammation play important roles in cancer development.

This book is the culmination of the efforts of several researchers, scientists, graduate students and post-doctoral fellows across the world. In this book authors focused on the role of natural compounds in the prevention and therapy of various cancers. The book has enormous scope and will benefit multiple audience including researchers, clinicians, patients, academicians, industrialists, and students. The editors are also thankful to Bentham Publisher and their team members for the opportunity to publish this book. Lastly we thank our family members for their love, support, encouragement and patience during the entire period of this work.

Sahdeo Prasad & Amit K Tyagi
Department of Experimental Therapeutics, Division of Cancer Medicine
The University of Texas M. D. Anderson Cancer Center
Houston, TX 77054, USA

List of Contributors

Amit K. Tyagi	Department of Experimental Therapeutics, Division of Cancer Medicine, The University of Texas M. D. Anderson Cancer Center, Houston, USA
Ammad Ahmad Farooqi	Institute of Biomedical and Genetic Engineering (IBGE), Islamabad, Pakistan
Arvind Kumar	Natural Product Microbes, CSIR-Indian Institute of Integrative Medicine (CSIR), Canal Road, Jammu Tawi, India
B.S. Dwarkanath	Institute of Nuclear Medicine and Allied Science, Street no., SK Majumdar Marg, India
Bhahwal Ali Shah	Natural Product Microbes, CSIR-Indian Institute of Integrative Medicine (CSIR), Canal Road, Jammu Tawi, India
Conrad V. Simoben	Pharmaceutical Chemistry, Martin-Luther Universität Halle-Wittenberg, Wolfgang-Langenbeck-Str. 4, 06120 Halle (Saale), Germany Chemistry Department, University of Buea, South West Region, Central Africa
Faiza Yasmeen	Institute of Blood Transfusion Services, Lahore, Pakistan
Fidele Ntie-Kang	Pharmaceutical Chemistry, Martin-Luther Universität Halle-Wittenberg, Wolfgang-Langenbeck-Str. 4, 06120 Halle (Saale), Germany Chemistry Department, University of Buea, South West Region, Central Africa
Ilhan Yaylim	Istanbul University, Department of Molecular Medicine, Institute of Experimental Medicine Istanbul, Istanbul, Turkey
Manjeet Kumar	Natural Product Microbes, CSIR-Indian Institute of Integrative Medicine (CSIR), Canal Road, Jammu Tawi, India
Mehak Gulzar	Department of Biotechnology, School of Engineering and Technology, Sharda University, Knowledge Park-III, Gautam Buddha Nagar, Street no., Greater Noida, India Nutrametrix Health Solutions, USA
Muhammad Zahid Qureshi	GCU Department of Chemistry, Lahore, Pakistan
Neetu Kumra Taneja	National Institute of Food Technology Entrepreneurship and Management, Kundli, India
Omkar P. Dhamale	Sanofi Pasteur, 1 Discovery Drive, Swiftwater, USA
Pankaj Taneja	Department of Biotechnology, School of Engineering and Technology, Sharda University, Knowledge Park-III, Gautam Buddha Nagar, Greater Noida, India Nutrametrix Health Solutions, USA
R. Jayaraj	Biochemistry Laboratory, Non-Timber Forest Produce Department, Kerala Forest Research Institute, Peechi, Thrissur, India
R.P. Tripathi	Institute of Nuclear Medicine and Allied Science, Street no., SK Majumdar Marg, India
Roshan Lal	Biochemistry Laboratory, Non-Timber Forest Produce Department, Kerala Forest Research Institute, Peechi, Thrissur, India
Rukset Attar	Yeditepe University Medical School Istanbul, Istanbul, Turkey
Sahdeo Prasad	Department of Experimental Therapeutics, Division of Cancer Medicine, The University of Texas M. D. Anderson Cancer Center, Houston, USA

Sandeep K. Misra	University of Mississippi, Oxford, USA 38677
Sanjay Mishra	Proteomics and Environmental Carcinogenesis Laboratory, CSIR-Indian Institute of Toxicology Research, M.G. Marg, Lucknow, India
Shankar Suman	Proteomics and Environmental Carcinogenesis Laboratory, CSIR-Indian Institute of Toxicology Research, M.G. Marg, Lucknow, India
Shinjini Singh	Cytokine Research Laboratory, Department of Experimental Therapeutics, The University of Texas, M.D. Anderson Cancer Center, Houston, USA
Sobia Tabassum	Department of Bioinformatics and Biotechnology, International Islamic University, Islamabad, Pakistan
Yogeshwer Shukla	Proteomics and Environmental Carcinogenesis Laboratory, CSIR-Indian Institute of Toxicology Research, M.G. Marg, Lucknow, India

Cancer Preventive and Therapeutic Compounds: Gift From Mother Nature

Cancer Preventive and Therapeutic Compounds: Gift From Mother Nature

2

CHAPTER 1

Dietary Agents: Effective and Safe Natural Assets Against Cancer

Sahdeo Prasad* and **Amit K. Tyagi**

Department of Experimental Therapeutics, Division of Cancer Medicine, The University of Texas M. D. Anderson Cancer Center, Houston, TX 77054, USA

Abstract: Cancer stands as the second most common cause of disease-related death in humans. Although numerous anticancer drugs are available, mostly they are expensive with serious side effects. Thus, the challenging task of finding an alternative cancer treatment measure has become more important than ever to both scientists and physicians. Since natural compounds are known for their various health benefits for centuries, several nutritional factors have brought considerable attention as modifiable risk factors in the prevention and treatment of cancer. Based on currently available research, the present chapter focuses on the chemo preventive and chemotherapeutic properties of different natural/dietary compounds such as fruits, vegetable, spices, nuts, legumes, cereals and grains highlighting their potential use against cancer treatment. The molecular mechanisms by which theses dietary compounds inhibit cancer development and induce cell death are also included to a certain extent.

Keywords: Cell signaling molecules, Cancer, Chemoprevention and chemotherapy, Inflammation, Natural compounds.

INTRODUCTION

Cancer is one of the major health problems not only in the United States, but also in many other countries of the world. At present, 1 in 3 women and 1 in 2 men in the USA will develop cancer in their lifespan [1]. Although radiation-therapy, surgery and chemotherapy are the standard ways to treat most kinds of cancers, reoccurrence of cancer and resistance to chemotherapeutic drugs have become major impediments for treating cancer. Moreover, multidrug resistance of cancer is now considered a main reason for the failure of chemotherapy [2]. Alternatives

_____

*** Corresponding author Sahdeo Prasad:** Department of Experimental Therapeutics, Division of Cancer Medicine, The University of Texas MD Anderson Cancer Center, Houston, TX 77054, USA; Tel: 713-792-6459; Fax: 713-745-1710; E-mails: spbiotech@gmail.com, amittyagiiitd@gmail.com

that are inexpensive, effective, and safe compared with synthetic chemo-therapeutic agents are profoundly required. An abundance of epidemiological, clinical and experimental evidences have shown that use of the compounds from natural sources play an important preventive and therapeutic role in the etiology of human cancers (Fig. **1**).

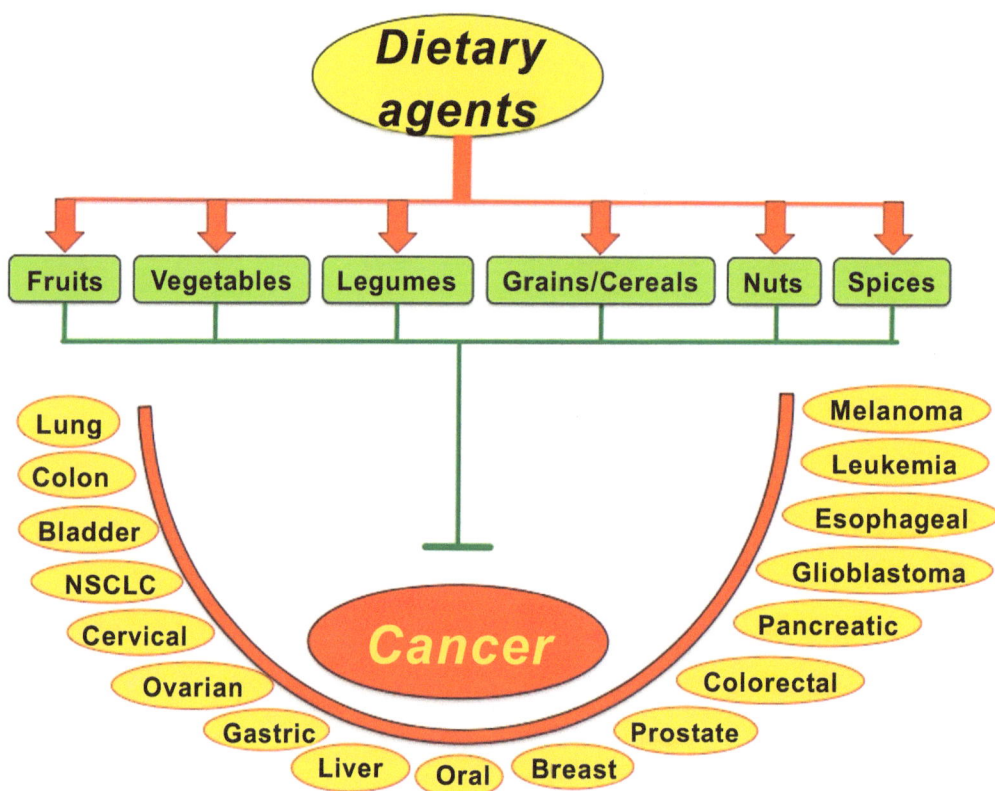

various diseases including cancer. Many drugs against cancer approved by FDA are originated from natural products or their derivatives. Still several natural compounds are under clinical trials. These natural products include fruits, vegetables, legumes, cereals, and nuts, which are being routinely consumed by human (Fig. **2**). The anticancer compounds are usually present in any part of the plant such as roots, leaf, bark, flower, fruits etc. Accumulated data from preclinical and clinical studies revealed that these natural compounds are chemically diverse because they can act at several stages of tumor development. In this chapter we discuss some selected type of cancer preventive and therapeutic compounds provided by Mother Nature.

Fig. (2). Selected cancer preventive and therapeutic compounds provided by Mother Nature.

FRUITS AND VEGETABLES

Varieties of phytochemicals are reported to found in fruits and vegetables, which could have potential against the cancer. These phytochemicals have antioxidant and anti-inflammatory activities, which contribute to the prevention of tumorigenesis. Because chronic inflammation and oxidative stress are associated with the development of cancer, fruits and vegetables with these properties hold promise for cancer chemoprevention. These phytochemicals in fruits and vegetables can help prevent cancer [3]. The phytochemicals present in fruits and vegetables modulate large number of cell signaling molecules those linked with cancer. The modulation of signaling molecules controls the abnormal growth of cells and ultimately controls the growth of cancer.

A large number of laboratory studies have revealed that phytochemicals from fruits and vegetables inhibit every step of cancer development and thus display chemo preventive and chemotherapeutic potential against cancer cells. These phytochemicals exhibit anticancer activity against varieties of cancers including lung carcinoma, oral cancer, leukemia, breast cancer, multiple myeloma, head and neck cancer, prostate cancer, pancreatic and other cancer. Ursolic acid, found in various fruits, inhibits growth and proliferation of colon, multiple myeloma, leukemia and others through multiple pathways [4]. Indole-3-carbinol from cruciferous vegetables also inhibits growth and reverses the multidrug resistance of leukemic cells [5]. Resveratrol suppresses survival, growth, proliferation, and metastasis of pancreatic cancer through inhibition of the inflammatory molecules such as NF-kappaB, STAT3 as well as expression of various other cell signal molecules including Bcl-2, cFLIP, cyclin D1, CXCR4, MMP-9, ICAM-1, and VEGF [6].

Date palm fruits (*Phoenix dactylifera* L.), which is a rich source of dietary fiber and polyphenols, has been found to inhibit Caco-2 cell growth. The study suggests that intake of date are helpful in the maintenance of bowel health also reduce the incidence of colorectal cancer in human [7]. Betulinic acid, present in sour jujube fruits, induces apoptosis in breast cancer MCF-7 cells. This indicates that consumption of sour jujube fruits may reduce the occurrence of breast cancer [8]. Strawberry fruit has shown anticancer activity in leukemia (CEM) and breast cancer (T47D) cell lines [9]. These data suggest that there are numerous other compounds from fruits and vegetables have efficacy against cancer cells.

Based on *in vitro* experiment, animal studies were conducted to validate whether fruits and vegetable are effective in the body systems. In mice models, strawberry fruit has shown cancer therapeutic and chemo preventive potentials against breast cancer. When breast adenocarcinoma bearing mice were treated with strawberry fruit extract inhibited the growth of cancer cells and resulted in an extended life [9]. When treating colorectal cancer, ursolic acid inhibited tumor growth and chemo sensitizes to chemotherapeutic drugs in a nude mice model [4]. In an orthotopic mouse model of human pancreatic cancer, resveratrol found to be inhibiting pancreatic tumor growth and this inhibitory effect of resveratrol was further increased by the treatment of gemcitabine [6]. γ-tocotrienol found in palm fruit not only inhibited growth of pancreatic tumor cells, but also inhibited tumor development and enhanced the gemcitabine-induced inhibition of human tumor cell growth in an orthotopic nude mouse model [10]. Lupeol, which is present in various fruits, inhibits skin tumor growth in mice [11] and hepatocellular carcinoma [12]. Another compound, luteolin, found in several fruits and vegetables, have shown potent antioxidant and anti-inflammatory activity. When luteolin was used to treat nude mice with human bronchial epithelial (BEAS-2B) cells, tumor incidence was reduced compared to a control group [13]. Thus numerous other data indicate that polyphenols present in various vegetables and fruits have the potential to fight cancer.

Epidemiological studies have also been examined to determine the relationship between intake of fruits and vegetables and incidence of various cancers. Low consumption of fruits and vegetables are linked with high risk of cancer as observed in both prospective and retrospective studies. In diagnosed cases of lung cancer of male smokers in China, it has been observed that intake of fruits and vegetables rich in vitamin A and/or carotenoids had lower incidence of cancer. However patients with low consumption of yellow and green vegetables had significantly increased lung cancer [14]. Another cohort study with retirement community initially had no cancer, some cases were found to be diagnosed with cancer. In women, reduced cancer risks were noted for those who ate all vegetables and fruits rich in vitamin C [15]. Fung *et al.* [16] also found that consumption of high amount of fruits such as berries and peaches leads to lower risk of estrogen receptor negative breast cancer in postmenopausal women.

In a population-based, prospective cohort study a strong link between consumption of fruits and vegetables with lung cancer risk were found. Furthermore, the intake of carotenoid-rich vegetables among 61,491 adult Chinese men is associated with lower risk of lung cancer [17]. Other prospective data from a cohort study done in the Netherlands showed a relationship between high intake of fruits and vegetables with lower risk of head-neck cancer. In a 20.3 year follow-up study, it was found that the 120,852 participants showed inverse association with fruit and vegetable intake and head and neck cancer risk. However, no significant association was found [18]. These studies indicate that the probably polyphenols present in the fruits and vegetables are fighting against cancer development.

The response of vegetables, fruits, and carotenoids with cancer development was also reported to be different with different races and sexes. In a study of smokers in United States, consumption of fruits, vegetables, and carotenoids by white females had significantly lower risk of lung cancer, while white males had a non-significant association, and black females showed an inverse association just with vegetables. However, associations between consumption of these dietary agents with association of lung cancer risk were not found in black males. This study indicates that intake of fruits and vegetables along with carotenoids may be helpful in the prevention of lung cancer in smokers [19]. Another study also revealed that consumption of higher amount of fruits and vegetables is associated with a modest reduction of primarily cardiovascular disease and not significant reduction of cancer [20]. However, routinely consumed fruits vegetables have not shown any potential carcinogenicity.

The reason behind this is a subject of investigation. Besides these, numerous other studies also found no association of cancer risk with consumption of fruits and vegetables. However, intake of fruits and vegetables, at least, did not adversely affect cancer risk. Conclusively, appropriate intake of fruits and vegetables could prevent the cancer risk without causing any side effects.

Moreover, fruits and vegetables have been tested clinically in cancer patients. For example, resveratrol against colorectal cancer (NCT00433576), hepatocellular carcinoma (NCT01476592); ursolic acid against benign prostate hyperplasia (NCT02702947); tocotrienol against lung cancer (NCT02644252, NCT-00002586), head and neck cancer (NCT00054561), bladder cancer (NCT-00553345, NCT00553124). Besides these several other natural compounds have used to treat cancer patients (https://clinicaltrials.gov). These compounds have shown potential against the cancer without any adverse side effects and recommended for their further use.

LEGUMES

In many regions of the world, legumes are considered as part of traditional diet. Accumulated studies revealed that legumes such as pulses have cancer preventive properties because of the polyphenols present in the legumes. The most widely studied phytochemicals isoflavones are found in soybeans, which has enormous activity against cancer. The isoflavones such as equol, daidzein, and genistein obtained from different legumes have cancer chemo preventive properties against wide varieties of cancer. Epidemiological studies revealed that consumption of isoflavone-rich soy-based diet is inversely associated with incidence and mortality from cancer. Besides these, lower reduced incidence rate and mortality of prostate cancer in Asian countries was observed due to high intake of soy isoflavones rich diet in their daily lifestyle. It has been also observed that Asian countries have lower incidence of breast cancer with the putative antiestrogenic effects of isoflavones [21]; however, a study conduced on Western population having low consumption of soy food, no association between soy isoflavones and cancer incidence was observed [22]. Further studies reveal that the age at which people are exposed to soy foods probably latter affect the incidence of cancer in those populations. It has also been found that consumption of high amount of soy during adolescence could reduce the risk of breast cancer incidence later in life [23, 24] however, consumption of soy late stage of life may not be effective in preventing breast cancer [22].

A case-control study conducted in Uruguay between the year 1996 and 2004, with 3,539 cancer patients and 2,032 control cases, a strong association between legume consumption and the cancer risk were observed. Consumption of high amount of legumes was found to be linked with a reduced risk of various cancers such as colorectum, stomach, upper aerodigestive tract and kidney cancer [25]. Numerous other studies showed that the regularly legumes and beans consuming population have reduced risk of colorectal adenomas. In a 6-year prospective study with 32,051 non-hispanic, white, cohort members, an inverse relation between bean consumption and incident of colon cancer was found. People who were consuming legumes at least two times in a week the risk of colon cancer incidence reduced to half [26]. How beans prevent the risk of cancer is not very much illustrated. However, numerous studies have shown that legumes target several cell signaling molecules. It has also been shown that the fiber and resistant starch present in the beans and other legumes pass into the large intestine, where they further ferment into short chain fatty acids like butyrate. This butyrate probably exhibits cancer-preventive actions in the colon [27, 28].

Experimental studies conducted by using cancer cells and animals models also revealed that legume component such as daidzein, genistein, equol and other display cancer chemo preventive activity through the regulation of cell growth and proliferative molecules. Daidzein, a soy flavonoid, inhibits cell growth by arresting cell cycle and inducing apoptosis of colorectal, cervical, ovarian and several other cancer cells [29, 30]. In animal model of cancer, daidzein and its metabolite equol exhibited potential in suppressing mammary tumors growth [31]. Genistein also shown chemo preventive and therapeutic potential against wide varieties of cancer by acting at several steps of cancer development such as cell survival, proliferation, metastasis as well as induction of apoptosis [32]. Genistein and its related isoflavones also inhibit growth and development of stomach, bladder, prostate, lung, and hematopoietic cancer induced by chemical carcinogens [33]. In addition, many experimental and epidemiological results with the mechanism of action revealed that these compounds have cancer chemo preventive and therapeutic properties. Thus, these evidences provide the opportunity to scientist around the world to explore the importance of these phytoestrogens present in legumes against cancer and other chronic diseases.

GRAINS AND CEREALS

Grains and cereals are routinely consumable diet, which include rice, wheat, corn, rye, oats, barley, millet etc. Consumption of grains and cereals are found to be linked with a reduced risk of various common diseases including cancer, obesity, heart disease and type 2 diabetes. Long back ago in 1930s the association of wholegrain food intake and cancer prevention has been studied UK population where they found that consumption of whole meal bread prevents the cancer risk [34]. In contrast, refined grains were proportionately associated with the cancer of various organs such as the oesophagus, larynx, oral cavity and pharynx [35]. However, many other findings suggested that cereals and grains containing high fiber reduce the cancer risk.

A case-control study conducted in Northern Italy further provided an evidence that high intake of wholegrain foods has strong relationship with reduced risk of cancer. The study revealed that wholegrain food consumption reduced the risk of several cancer in those include cancer of lungs, liver, mouth, kidney, colon, rectum, pancreas, and blood [36]. Gil [37] also reported that cancers of digestive tract, hormonal, and pancreatic cancer could be prevented by regular intake of wholegrain cereals and foods. Recently, in a case-control study with newly diagnosed breast cancer female patients, it has been found that consumption of wholegrain food more than seven times per week was directly linked with a 0.49-fold reduced risk of breast cancer [38]. In another study, it has been found that intake of wholegrain rye bread, whole-wheat bread and oatmeal, during different periods of life, was linked with the prostate cancer risk, particularly advanced disease [39].

Laboratory studies further disclose the mechanism of anticancer properties of cereals and grains. Barley, a commonly used cereal, exhibited anti-proliferative activity on human leukemia/lymphoma cell lines. Barley extract has shown to cause cell cycle arrest and further cell death mediated by cleavage of caspases and PARP [40]. Another cereal known as millet, exhibited anticancer effects by suppressing proliferation of human colorectal cancer cell lines. The anticancer effect of millet was through preferential ROS accumulation in cancer cells, which appeared to the suppression of NF-E2-related factor 2 (Nrf2), as well as decreased activity of antioxidant enzyme catalase and glutathione [41]. Phytosterols, particularly oxyphytosterols, present in wheat bran have proven to possess antiproliferative properties of colon cancer cells, and thereby contributed to the chemoprevention of whole grain wheat [42]. In addition to these, Jayaram *et al* [43] showed that pectic polysaccharide present in corn suppressed growth and metastasis of cancer cells by acting at multiple steps of cancer development. This anticancer and antimetastasis property of corn was mediated through regulation of VEGF matrix metalloproteinases 2 and 9 and NF-κB [43].

In animal studies, cereal and grains were found to be effective against cancer prevention. Lignan from wheat bran reported to prevent colon cancer in APC-Min mice model. This lignan metabolites-induced suppression of colon cancer proliferation was mediated through modulation of cytostatic and apoptotic pathways [44]. Millet exhibited an *in vivo* anti-tumor effect by suppressing xenografted tumor growth in nude mice [45]. In APC-Min mice model of colon cancer, intake of rice bran reduced the burden of adenoma in animals [46]. Along with these, there are numerous other findings suggesting that intake of grain and wheat is directly linked with lower cancer risk. However, numerous other studies also revealed nonsignificant association.

NUTS

Nuts contain high amounts of polyphenols, phytochemicals that have antioxidant, anti-inflammatory properties. These properties potentially participate in protecting from cancer risk. Epidemiological studies suggested a consumption of nuts are directly associated with cause-specific mortality. In a cohort study it has been observed that consumption of nuts are potentially associated with total mortality and mortality due to cancer and cardiovascular diseases [47]. In a case-control study on patients of stomach cancer, consumption of dietary foods including nuts were found to be associated with reduced risk of cancer with dose dependent response [48]. A data obtained from another cohort study with 14000 Adventist men showed that the intake of high amount of fruits, vegetables, beans and nuts statistically reduce the incidence of prostate cancer [49]. Nuts also play a chemo preventive role against cancer in female, which is evident by a prospective study conducted in Tiawan. Observation indicates that consumption of nuts prevents the risk of colorectal cancer in women [50]. Besides these, 27% reduced risk of endometrial cancer was found in Greek women those who were consuming diet rich in nuts, seeds and legumes [51]. Taken together, these studies indicate that intake of nuts are inversely associated with cancer.

In vitro and animal studies also suggest that nuts have preventive effects against tumor growth and proliferation. Lux *et al.* [52] showed that supernatant obtained from fermentation of some of the edible nuts (almonds, macadamias, hazelnuts, pistachios, walnuts) suppress the proliferation of colorectal cancer HT29 cells also exhibit anti-genotoxic activity. These nuts extracts showed strong antioxidant potential and reduced H_2O_2-induced DNA damage [52]. Dietary walnuts also suppress colorectal cancer growth in animals. When a diet containing walnuts was given to mice with xenografted human HT-29 colon tumor cells, a significant suppression in tumor development was detected. Walnut-induced inhibition of colorectal cancer growth was mediated by suppressing angiogenesis [53]. Ethanol extract of walnuts induced cytotoxicity to human breast tumor MCF-7 and MDA-MB-231, as well as HeLa cancer cells. The compounds those induced cytotoxicity were identified as Tellimagrandin I and Tellimagrandin II, which was found to be ellagitannin family members. A diet supplemented with walnuts has also shown to decrease the size of human MDA-MB-231 tumor transplanted in nude mice [54]. All these studies indicate that nuts are effective in suppressing human tumor growth either in cell culture, animals or in humans.

SPICES

Since centuries spices have been used for different medicinal purposes including to treat stomach upset, inflammation, infection, wound healing etc in Indian Ayurvedic medicine. However, few decades ago spices explored as an anticancer agent in experimental studies. Spices have the ability to modulate various steps of cell signaling pathways, which are linked to the tumorigenesis. They can inhibit transformation of cells, cancer cell survival, activation of carcinogens, oxidative stress, inflammation and enhance apoptosis of tumor cells. Spices have multi-targeted nature with known mechanisms by which they may serve as anticancer, beat the diseases and boost the health. Till today over 180 bioactive compounds from spices have been identified and investigated for their health benefits to human beings [55]. In this chapter we will deal some of the major spices for their anticancer and health benefit effects.

One of the spices is turmeric, which is widely documented as potent anticancer agent in experimental as well as clinical studies. Curcumin, an active component of turmeric has shown effective as an anticancer agent in clinical trials. Based on over 50 clinical trials using human subjects, curcumin has been found to be both safe and effective against several diseases and disorders including cancer, and other chronic diseases [56]. Garlic also displayed anticancer effects as evidenced by numerous clinical studies. In a clinical trial (RCT, level II), the patients with colon cancer taking aged garlic extract experienced 29% reduction in size and number of adenomas compared to the placebo. In other cohort studies (level II) 5 of 8 cases experienced protective effect by consumption of raw or cooked garlic, however a protective effect for distal colon was observed in 2 of 8 cases by intake of garlic. Meta-analysis (level III) data showed that an inverse relationship with risk of cancer, where consumption of garlic reduced cancer risk by 30% [57]. Besides these, ginger has also reported to be cancer preventive and therapeutic. Recently, we have summarized that ginger and its active components prevent the risk of gastrointestinal cancer. The anticancer properties of ginger was associated with its antioxidant, anti-inflammatory and other biological properties [58]. These studies indicate that spices have the ability to against cancer in humans.

As reported previously, spices are full of various types of antioxidants and anti-inflammatory components. These components mainly include various phenols and flavonoids. In an experimental study with 9 spices (including ginger, caraway, cumin, fennel, pepper, long pepper and others), it has been found that these spices protected DNA damage caused by H_2O_2 and nicotine in 3T3-L1 (mouse fibroblasts) and MCF-7 (breast cancer) cells [59]. Several other spices and traditionally used medicinal plants such as *Fagara leprieuri, Fagara xanthoxyloïdes, Mondia whitei* and *Xylopia aethiopica*, have efficacy against cancer. The extracts of these spices were cytotoxic to the breast cancer MCF-7 cell line [60]. Nutraceuticals derived from spices have been reported to suppress growth of different types of cancer cells through the modulation of multiple pathways including inflammatory NF-κB signaling pathway. Curcumin, diosgenin, gambogic acid, capsaicin, ursolic acid, noscapine, sesamin, anethole and eugenol, which are the active components of spices, have shown to inhibit NF-κB [61]. Besides these, various other spice-derived nutraceuticals shon in Fig. (**2**) have chemo preventive and chemotherapeutic properties. These spices exhibit anticancer properties through the modulation of various inflammatory mediators, growth factors, transcription factors, and protein kinases. Through these studies, it has become evident that spice and spice-derived nutraceuticals have efficacy to prevent the incidence of cancer by modulating various cell-signaling pathways.

Spices have been used with or without conventional drugs for the prevention and treatment of cancer in human. Curcumin was clinically used for the prevention of gastric cancer (NCT02782949), colon cancer (NCT00027495). It was used for the treatment of bone cancer (NCT00689195), breast cancer (NCT02556632), head and neck cancer (NCT01160302) and several other cancers. Other spices including sesame to treat skin reactions from radiotherapy of breast cancer (NCT01688479) have been used (https://clinicaltrials.gov). Several other natural compounds are under the consideration for clinical trial against multiple types of cancer.

CONCLUSION

It is well known that dysregulation of multiple cell-signaling pathways are involved in cancer. Accumulated evidences show that dietary agents can modulate these pathways effectively indicating their enormous potential against cancer. As per the evidences mentioned above, dietary nutraceuticals are fascinating because they exhibit anti-cancer activity without affecting normal cells, thus they warrant greater attention. Instead of it, dietary nutraceuticals have high efficacy against multiple diseases (Fig. **3**). They are orally bio-available, have a known mechanism of action, easily accessible, cost effective, and are acceptable to the all region of human population. However, very few clinical studies have been performed so far. Therefore, more clinic studies are required to validate the experimental findings and answer some of unresolved questions, including safety, bioavailability, molecular targets efficacy of these natural dietary compounds before the use of nutraceuticals as part of cancer prevention or a treatment regimen.

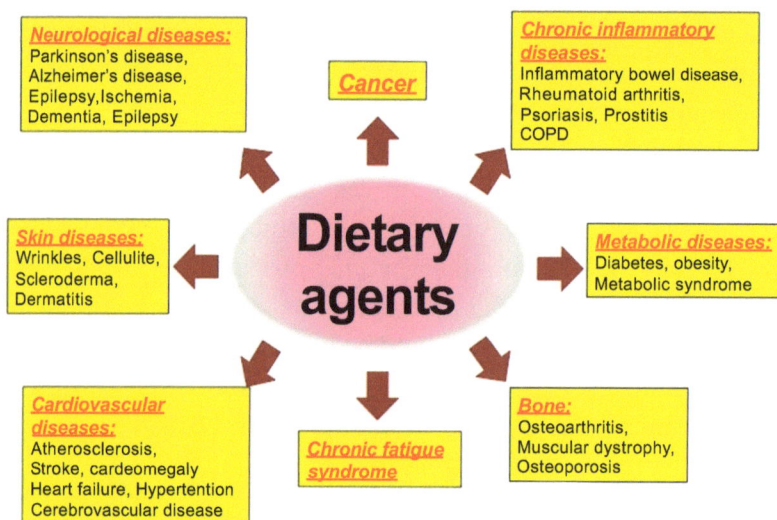

Fig. (3). Beneficial effects of dietary compounds against various diseases.

CONFLICT OF INTEREST

The authors confirm that they have no conflict of interest to declare for this publication.

ACKNOWLEDGEMENTS

Declared none.

REFERENCES

[1]   Siegel RL, Miller KD, Jemal A. Cancer statistics, 2016. CA Cancer J Clin 2016; 66(1): 7-30.
      [http://dx.doi.org/10.3322/caac.21332] [PMID: 26742998]

[2]   Singh N, Sidiq Z, Bhalla M, Myneedu VP, Sarin R. Multi-drug resistant tuberculosis among category I treatment failuresa retrospective study. Indian J Tuberc 2014; 61(2): 148-51.
      [PMID: 25509938]

[3]   Kundu JK, Chun KS. The promise of dried fruits in cancer chemoprevention. Asian Pac J Cancer Prev 2014; 15(8): 3343-52.
[http://dx.doi.org/10.7314/APJCP.2014.15.8.3343] [PMID: 24870720]

[4]   Prasad S, Yadav VR, Kannappan R, Aggarwal BB. Ursolic acid, a pentacyclin triterpene, potentiates TRAIL-induced apoptosis through p53-independent up-regulation of death receptors: evidence for the role of reactive oxygen species and JNK. J Biol Chem 2011; 286(7): 5546-57.
[http://dx.doi.org/10.1074/jbc.M110.183699] [PMID: 21156789]

[5]   Arora A, Seth K, Kalra N, Shukla Y. Modulation of P-glycoprotein-mediated multidrug resistance in K562 leukemic cells by indole-3-carbinol. Toxicol Appl Pharmacol 2005; 202(3): 237-43.
[http://dx.doi.org/10.1016/j.taap.2004.06.017] [PMID: 15667829]

[6]   Harikumar KB, Kunnumakkara AB, Sethi G, *et al.* Resveratrol, a multitargeted agent, can enhance antitumor activity of gemcitabine *in vitro* and in orthotopic mouse model of human pancreatic cancer. Int J Cancer 2010; 127(2): 257-68.
[PMID: 19908231]

[7]   Eid N, Enani S, Walton G, *et al.* The impact of date palm fruits and their component polyphenols, on gut microbial ecology, bacterial metabolites and colon cancer cell proliferation. J Nutr Sci 2014; 3: e46.
[http://dx.doi.org/10.1017/jns.2014.16] [PMID: 26101614]

[8]   Sun YF, Song CK, Viernstein H, Unger F, Liang ZS. Apoptosis of human breast cancer cells induced by microencapsulated betulinic acid from sour jujube fruits through the mitochondria transduction pathway. Food Chem 2013; 138(2-3): 1998-2007.
[http://dx.doi.org/10.1016/j.foodchem.2012.10.079] [PMID: 23411336]

[9]   Somasagara RR, Hegde M, Chiruvella KK, Musini A, Choudhary B, Raghavan SC. Extracts of strawberry fruits induce intrinsic pathway of apoptosis in breast cancer cells and inhibits tumor progression in mice. PLoS One 2012; 7(10): e47021.
[http://dx.doi.org/10.1371/journal.pone.0047021] [PMID: 23071702]

[10]   Kunnumakkara AB, Sung B, Ravindran J, *et al.* Gamma-tocotrienol inhibits pancreatic tumors and sensitizes them to gemcitabine treatment by modulating the inflammatory microenvironment. Cancer Res 2010; 70(21): 8695-705.
[http://dx.doi.org/10.1158/0008-5472.CAN-10-2318] [PMID: 20864511]

[11]   Nigam N, Prasad S, George J, Shukla Y. Lupeol induces p53 and cyclin-B-mediated G2/M arrest and targets apoptosis through activation of caspase in mouse skin. Biochem Biophys Res Commun 2009; 381(2): 253-8.
[http://dx.doi.org/10.1016/j.bbrc.2009.02.033] [PMID: 19232320]

[12]   He Y, Liu F, Zhang L, *et al.* Growth inhibition and apoptosis induced by lupeol, a dietary triterpene, in human hepatocellular carcinoma cells. Biol Pharm Bull 2011; 34(4): 517-22.
[http://dx.doi.org/10.1248/bpb.34.517] [PMID: 21467639]

[13]   Pratheeshkumar P, Son YO, Divya SP, *et al.* Luteolin inhibits Cr(VI)-induced malignant cell transformation of human lung epithelial cells by targeting ROS mediated multiple cell signaling pathways. Toxicol Appl Pharmacol 2014; 281(2): 230-41.
[http://dx.doi.org/10.1016/j.taap.2014.10.008] [PMID: 25448439]

[14]   Forman MR, Yao SX, Graubard BI, *et al.* The effect of dietary intake of fruits and vegetables on the odds ratio of lung cancer among Yunnan tin miners. Int J Epidemiol 1992; 21(3): 437-41.
[http://dx.doi.org/10.1093/ije/21.3.437] [PMID: 1634303]

[15]   Shibata A, Paganini-Hill A, Ross RK, Henderson BE. Intake of vegetables, fruits, beta-carotene, vitamin C and vitamin supplements and cancer incidence among the elderly: a prospective study. Br J Cancer 1992; 66(4): 673-9.

[http://dx.doi.org/10.1038/bjc.1992.336] [PMID: 1419605]

[16]   Fung TT, Chiuve SE, Willett WC, Hankinson SE, Hu FB, Holmes MD. Intake of specific fruits and vegetables in relation to risk of estrogen receptor-negative breast cancer among postmenopausal women. Breast Cancer Res Treat 2013; 138(3): 925-30.
[http://dx.doi.org/10.1007/s10549-013-2484-3] [PMID: 23532538]

[17]   Takata Y, Xiang YB, Yang G, *et al.* Intakes of fruits, vegetables, and related vitamins and lung cancer risk: results from the Shanghai Mens Health Study (20022009). Nutr Cancer 2013; 65(1): 51-61.
[http://dx.doi.org/10.1080/01635581.2013.741757] [PMID: 23368913]

[18]   Maasland DH, van den Brandt PA, Kremer B, Goldbohm RA, Schouten LJ. Consumption of vegetables and fruits and risk of subtypes of head-neck cancer in the Netherlands Cohort Study. Int J Cancer 2015; 136(5): E396-409.
[http://dx.doi.org/10.1002/ijc.29219] [PMID: 25220255]

[19]   Dorgan JF, Ziegler RG, Schoenberg JB, *et al.* Race and sex differences in associations of vegetables, fruits, and carotenoids with lung cancer risk in New Jersey (United States). Cancer Causes Control 1993; 4(3): 273-81.
[PMID: 8318643]

[20]   Hung HC, Joshipura KJ, Jiang R, *et al.* Fruit and vegetable intake and risk of major chronic disease. J Natl Cancer Inst 2004; 96(21): 1577-84.
[http://dx.doi.org/10.1093/jnci/djh296] [PMID: 15523086]

[21]   Messina MJ. Legumes and soybeans: overview of their nutritional profiles and health effects. Am J Clin Nutr 1999; 70(3) (Suppl.): 439S-50S.
[PMID: 10479216]

[22]   Wu AH, Yu MC, Tseng CC, Pike MC. Epidemiology of soy exposures and breast cancer risk. Br J Cancer 2008; 98(1): 9-14.
[http://dx.doi.org/10.1038/sj.bjc.6604145] [PMID: 18182974]

[23]   Wu AH, Wan P, Hankin J, Tseng CC, Yu MC, Pike MC. Adolescent and adult soy intake and risk of breast cancer in Asian-Americans. Carcinogenesis 2002; 23(9): 1491-6.
[http://dx.doi.org/10.1093/carcin/23.9.1491] [PMID: 12189192]

[24]   Shu XO, Jin F, Dai Q, *et al.* Soyfood intake during adolescence and subsequent risk of breast cancer among Chinese women. Cancer Epidemiol Biomarkers Prev 2001; 10(5): 483-8.
[PMID: 11352858]

[25]   Aune D, De Stefani E, Ronco A, *et al.* Legume intake and the risk of cancer: a multisite case-control study in Uruguay. Cancer Causes Control 2009; 20(9): 1605-15.
[http://dx.doi.org/10.1007/s10552-009-9406-z] [PMID: 19653110]

[26]   Singh PN, Fraser GE. Dietary risk factors for colon cancer in a low-risk population. Am J Epidemiol 1998; 148(8): 761-74.
[http://dx.doi.org/10.1093/oxfordjournals.aje.a009697] [PMID: 9786231]

[27]   Williams EA, Coxhead JM, Mathers JC. Anti-cancer effects of butyrate: use of micro-array technology to investigate mechanisms. Proc Nutr Soc 2003; 62(1): 107-15.
[http://dx.doi.org/10.1079/PNS2002230] [PMID: 12740065]

[28]   Faris MA, Takruri HR, Shomaf MS, Bustanji YK. Chemopreventive effect of raw and cooked lentils (Lens culinaris L) and soybeans (Glycine max) against azoxymethane-induced aberrant crypt foci. Nutr Res 2009; 29(5): 355-62.
[http://dx.doi.org/10.1016/j.nutres.2009.05.005] [PMID: 19555818]

[29]   Guo JM, Xiao BX, Liu DH, *et al.* Biphasic effect of daidzein on cell growth of human colon cancer cells. Food Chem Toxicol 2004; 42(10): 1641-6.

[http://dx.doi.org/10.1016/j.fct.2004.06.001] [PMID: 15304310]

[30]   Hwang KA, Choi KC. Anticarcinogenic Effects of Dietary Phytoestrogens and Their Chemopreventive Mechanisms. Nutr Cancer 2015; 67(5): 796-803. [http://dx.doi.org/10.1080/01635581.2015.1040516] [PMID: 25996655]

[31]   Liu X, Suzuki N, Santosh Laxmi YR, Okamoto Y, Shibutani S. Anti-breast cancer potential of daidzein in rodents. Life Sci 2012; 91(11-12): 415-9. [http://dx.doi.org/10.1016/j.lfs.2012.08.022] [PMID: 23227466]

[32]   Spagnuolo C, Russo GL, Orhan IE, *et al.* Genistein and cancer: current status, challenges, and future directions. Adv Nutr 2015; 6(4): 408-19. [http://dx.doi.org/10.3945/an.114.008052] [PMID: 26178025]

[33]   Sarkar FH, Adsule S, Padhye S, Kulkarni S, Li Y. The role of genistein and synthetic derivatives of isoflavone in cancer prevention and therapy. Mini Rev Med Chem 2006; 6(4): 401-7. [http://dx.doi.org/10.2174/138955706776361439] [PMID: 16613577]

[34]   McIntosh G. Cereal foods, fibres and the prevention of cancers. Aust J Nutr Diet 2001; 58: 35-48.

[35]   Levi F, Pasche C, Lucchini F, Chatenoud L, Jacobs DR Jr, La Vecchia C. Refined and whole grain cereals and the risk of oral, oesophageal and laryngeal cancer. Eur J Clin Nutr 2000; 54(6): 487-9. [http://dx.doi.org/10.1038/sj.ejcn.1601043] [PMID: 10878650]

[36]   La Vecchia C, Chatenoud L, Negri E, Franceschi S. Session: whole cereal grains, fibre and human cancer wholegrain cereals and cancer in Italy. Proc Nutr Soc 2003; 62(1): 45-9. [http://dx.doi.org/10.1079/PNS2002235] [PMID: 12740056]

[37]   Gil A, Ortega RM, Maldonado J. Wholegrain cereals and bread: a duet of the Mediterranean diet for the prevention of chronic diseases. Public Health Nutr 2011; 14(12A): 2316-22. [http://dx.doi.org/10.1017/S1368980011002576] [PMID: 22166190]

[38]   Mourouti N, Kontogianni MD, Papavagelis C, *et al.* Whole grain consumption and breast cancer: a case-control study in women. J Am Coll Nutr 2016; 35(2): 143-9. [http://dx.doi.org/10.1080/07315724.2014.963899] [PMID: 25915188]

[39]   Torfadottir JE, Valdimarsdottir UA, Mucci L, *et al.* Rye bread consumption in early life and reduced risk of advanced prostate cancer. Cancer Causes Control 2012; 23(6): 941-50. [http://dx.doi.org/10.1007/s10552-012-9965-2] [PMID: 22527172]

[40]   Robles-Escajeda E, Lerma D, Nyakeriga AM, *et al.* Searching in mother nature for anti-cancer activity: anti-proliferative and pro-apoptotic effect elicited by green barley on leukemia/lymphoma cells. PLoS One 2013; 8(9): e73508. [http://dx.doi.org/10.1371/journal.pone.0073508] [PMID: 24039967]

[41]   Shan S, Shi J, Li Z, *et al.* Targeted anti-colon cancer activities of a millet bran-derived peroxidase were mediated by elevated ROS generation. Food Funct 2015; 6(7): 2331-8. [http://dx.doi.org/10.1039/C5FO00260E] [PMID: 26075747]

[42]   Zhu Y, Soroka D, Sang S. Oxyphytosterols as active ingredients in wheat bran suppress human colon cancer cell growth: identification, chemical synthesis, and biological evaluation. J Agric Food Chem 2015; 63(8): 2264-76. [http://dx.doi.org/10.1021/jf506361r] [PMID: 25658220]

[43]   Jayaram S, Kapoor S, Dharmesh SM. Pectic polysaccharide from corn (*Zea mays* L.) effectively inhibited multi-step mediated cancer cell growth and metastasis. Chem Biol Interact 2015; 235: 63-75. [http://dx.doi.org/10.1016/j.cbi.2015.04.008] [PMID: 25882088]

[44]   Qu H, Madl RL, Takemoto DJ, Baybutt RC, Wang W. Lignans are involved in the antitumor activity of wheat bran in colon cancer SW480 cells. J Nutr 2005; 135(3): 598-602. [PMID: 15735100]

[45] Shan S, Li Z, Newton IP, Zhao C, Li Z, Guo M. A novel protein extracted from foxtail millet bran displays anti-carcinogenic effects in human colon cancer cells. Toxicol Lett 2014; 227(2): 129-38. [http://dx.doi.org/10.1016/j.toxlet.2014.03.008] [PMID: 24685566]

[46] Norris L, Malkar A, Horner-Glister E, *et al.* Search for novel circulating cancer chemopreventive biomarkers of dietary rice bran intervention in Apc(Min) mice model of colorectal carcinogenesis, using proteomic and metabolic profiling strategies. Mol Nutr Food Res 2015; 59(9): 1827-36. [http://dx.doi.org/10.1002/mnfr.201400818] [PMID: 26033951]

[47] Guasch-Ferré M, Bulló M, Martínez-González MA, *et al.* Frequency of nut consumption and mortality risk in the PREDIMED nutrition intervention trial. BMC Med 2013; 11: 164. [http://dx.doi.org/10.1186/1741-7015-11-164] [PMID: 23866098]

[48] Hoshiyama Y, Sasaba T. A case-control study of stomach cancer and its relation to diet, cigarettes, and alcohol consumption in Saitama Prefecture, Japan. Cancer Causes Control 1992; 3(5): 441-8. [http://dx.doi.org/10.1007/BF00051357] [PMID: 1525325]

[49] Mills PK, Beeson WL, Phillips RL, Fraser GE. Cohort study of diet, lifestyle, and prostate cancer in Adventist men. Cancer 1989; 64(3): 598-604. [http://dx.doi.org/10.1002/1097-0142(19890801)64:3<598::AID-CNCR2820640306>3.0.CO;2-6] [PMID: 2743254]

[50] Yeh CC, You SL, Chen CJ, Sung FC. Peanut consumption and reduced risk of colorectal cancer in women: a prospective study in Taiwan. World J Gastroenterol 2006; 12(2): 222-7. [http://dx.doi.org/10.3748/wjg.v12.i2.222] [PMID: 16482621]

[51] Petridou E, Kedikoglou S, Koukoulomatis P, Dessypris N, Trichopoulos D. Diet in relation to endometrial cancer risk: a case-control study in Greece. Nutr Cancer 2002; 44(1): 16-22. [http://dx.doi.org/10.1207/S15327914NC441_3] [PMID: 12672637]

[52] Lux S, Scharlau D, Schlörmann W, Birringer M, Glei M. *In vitro* fermented nuts exhibit chemopreventive effects in HT29 colon cancer cells. Br J Nutr 2012; 108(7): 1177-86. [http://dx.doi.org/10.1017/S0007114511006647] [PMID: 22172380]

[53] Nagel JM, Brinkoetter M, Magkos F, *et al.* Dietary walnuts inhibit colorectal cancer growth in mice by suppressing angiogenesis. Nutrition 2012; 28(1): 67-75. [http://dx.doi.org/10.1016/j.nut.2011.03.004] [PMID: 21795022]

[54] Le V, Esposito D, Grace MH, *et al.* Cytotoxic effects of ellagitannins isolated from walnuts in human cancer cells. Nutr Cancer 2014; 66(8): 1304-14. [http://dx.doi.org/10.1080/01635581.2014.956246] [PMID: 25264855]

[55] Aggarwal BB, Kunnumakkara AB, Harikumar KB, Tharakan ST, Sung B, Anand P. Potential of spice-derived phytochemicals for cancer prevention. Planta Med 2008; 74(13): 1560-9. [http://dx.doi.org/10.1055/s-2008-1074578] [PMID: 18612945]

[56] Gupta SC, Sung B, Kim JH, Prasad S, Li S, Aggarwal BB. Multitargeting by turmeric, the golden spice: From kitchen to clinic. Mol Nutr Food Res 2013; 57(9): 1510-28. [http://dx.doi.org/10.1002/mnfr.201100741] [PMID: 22887802]

[57] Ngo SN, Williams DB, Cobiac L, Head RJ. Does garlic reduce risk of colorectal cancer? A systematic review. J Nutr 2007; 137(10): 2264-9. [PMID: 17885009]

[58] Prasad S, Tyagi AK. Ginger and its constituents: role in prevention and treatment of gastrointestinal cancer. Gastroenterol Res Pract 2015; 2015 [http://dx.doi.org/10.1155/2015/142979] [PMID: 142979]

[59]   Jayakumar R, Kanthimathi MS. Dietary spices protect against hydrogen peroxide-induced DNA damage and inhibit nicotine-induced cancer cell migration. Food Chem 2012; 134(3): 1580-4.
[http://dx.doi.org/10.1016/j.foodchem.2012.03.101] [PMID: 25005983]

[60]   Choumessi AT, Loureiro R, Silva AM, *et al.* Toxicity evaluation of some traditional African spices on breast cancer cells and isolated rat hepatic mitochondria. Food Chem Toxicol 2012; 50(11): 4199-208.
[http://dx.doi.org/10.1016/j.fct.2012.08.008] [PMID: 22902826]

[61]   Sung B, Prasad S, Yadav VR, Lavasanifar A, Aggarwal BB. Cancer and diet: How are they related? Free Radic Res 2011; 45(8): 864-79.
[http://dx.doi.org/10.3109/10715762.2011.582869] [PMID: 21651450]

CHAPTER 2

Medicinal Importance of Allicin – A Bioactive Component from *Allium Sativum* L (Garlic)

R. Jayaraj[*] and Roshan Lal

Biochemistry Laboratory, Non-Timber Forest Produce Department, Kerala Forest Research Institute, Peechi, Thrissur, Kerala – 680653, India

Abstract: *Allium sativum* L (garlic) has a lengthy history as being a food and spice having a unique taste and odor along with many medicinal properties. Garlic is considered to be a natural medicine against variety of human ailments, including various antibacterial, antiviral and antifungal infections, antithrombotic, anticancer and anti tumorogenic activities. All these activities are linked to the level of organosulfur compounds like allicin, flavonoids, and phenolic components in it. Freshly chopped garlic contains Allicin, which is one of the highly biologically dynamic component. Allicin has reported to have a number of bioactivities including antioxidant, anti-inflammatory activities. Many cardiovascular activities of allicin also have been worked out. The present paper reviews one of the major active ingredients in garlic – allicin – for its medicinal importance.

Keywords: Allicin, *Allium sativum*, Anti-cancer activities, Molecular targets.

INTRODUCTION

Dietary garlic is reported to have a number of health effects and is utilized over centuries for protection against microbial infections, chronic health effects and cancer. The medicinal properties of garlic (*Allium sativum*) has been known to mankind over thousands of years. The olden literatures and folklores of Egyptians, Indians, Romans, Babylonians and Greeks has mentioned the frequent use of garlic for various ailments such as respiratory infections, intestinal disorders, skin diseases, flatulence, worms, wounds, symptoms of aging are few among others [1].

The wound healing properties of garlic has been used by soldiers during World War II [2]. The spread of infections in wounds were inhibited through the

---

[*] **Corresponding author R. Jayaraj:** Biochemistry Laboratory, Non-Timber Forest Produce Department, Kerala Forest Research Institute, Peechi, Thrissur, Kerala – 680653, India; Tel/Fax: +91-487-2690147; Email: jayaraj@kfri.res.in

application of garlic paste directly to wounds. In addition to this, garlic displays a number of bioactivities such as procirculatory effects, hypolipidemic, and antiplatelet activities. The immune enhancement property of garlic is found to preclude flu and cold symptoms and also exhibits chemo preventive and anticancer activities, enhanced xenobiotic effects, regaining of physical fitness, radioprotection, anti-aging effects and stress reducing effects. Clinical and experimental studies on the intake of garlic preparations especially of garlic extract have shown widespread biological activities [3]. The garlic extracts reported to have hepatoprotective, neuroprotective, and antioxidative activities [4]. The *in vitro* activity of garlic against *Mycobacterium tuberculosis* has been reported long ago [5]. The cholesterol lowering effects of garlic supplements in humans were established in several clinical studies [6, 7]. According to the Ayurvedic and Greek systems of medicine, garlic is established as one of the best remedies for treatment of tuberculosis [8]. Many recent studies established that, consumption of allium-containing diet, lowered the threat of developing numerous malignancies [9 - 12], however the underlying signal transduction mechanisms were yet to be identified.

Both freshly chopped and crushed garlic is abundant with a group of organosulfur compounds called thiosulfinates, which are considered to be the reason for the beneficial effects. Garlic socked with warm oil contained mainly vinyldithiins, ajoene and small amount of sulfides [13, 14]. Freshly crushed garlic contains numerous sulfur compounds together labelled as garlic organosulfur compounds, which are the primary chemical entities accountable for the bioactivity of garlic. The major organosulfur chemical in garlic is (+) S-allyl-L-cysteinesulfoxide (alliin) that is normally stored away from the enzyme alliinase (EC 4.4.1.3). On crushing or chewing garlic, alliinase interacts with its substrate alliin to form 2-propenesulfenic acid, and on condensation it forms diallylthiosulfinate (allicin) [15]. Commercially obtainable garlic supplements are characterized into; garlic oil, dehydrated garlic powder, aged garlic extract (AGE) and garlic oil macerate.

Allicin - diallylthiosulfinate is one of the key bioactive compounds of garlic first isolated in 1944 and reported to have antifungal and antibacterial properties [16]. Allicin is reported to have enormous spectrum of health beneficial effects including: antihypertensive, cardioprotective, antimicrobial, antifungal, car-dioprotective, antiparasitic, anticancer and antiinflammatory activities [17 - 20].

Chemistry, Biosynthesis and Degradation

Allicin is chemically diallylthiosulfinate (IUPAC name: 3-prop-2-enylsulfinylsulfanylprop-1-ene) with a molecular formula $C_6H_{10}OS_2$ and molecular weight - 162.273 g/mol. Garlic (*Allium sativum*) is the natural source of allicin and garlic is being used as a food ingredient as well as in folk medicine across many civilization around the world since centuries. The defense mechanism of garlic against attacks by pests is through allicin. When the garlic plant is injured or attacked, it produces allicin by an enzymatic reaction and allicin is toxic to insects and microorganisms. The aroma of the freshly chopped garlic is also due to the presence of allicin [21]. Allicin is highly unstable and rapidly changes into a series of other sulfur containing compounds such as diallyl disulfide [22].

The enzyme alliinase acts on alliin (S-allyl-Lcysteine sulfoxide) forms allicin, which is present in the racemic form. Oxidation of diallyl disulfide also lead to the generation of racemic form of alliin [23]. In garlic cloves, alliin and alliinase are stored in different compartments, upon crushing of the cloves, both interact to form allicin, pyruvic acid and ammonia. Allylsulfenic acid is highly unstable and very reactive at room temperature. With the elimination of water, two molecules of allylsulfenic acid condense spontaneously to form allicin (Fig. **1**).

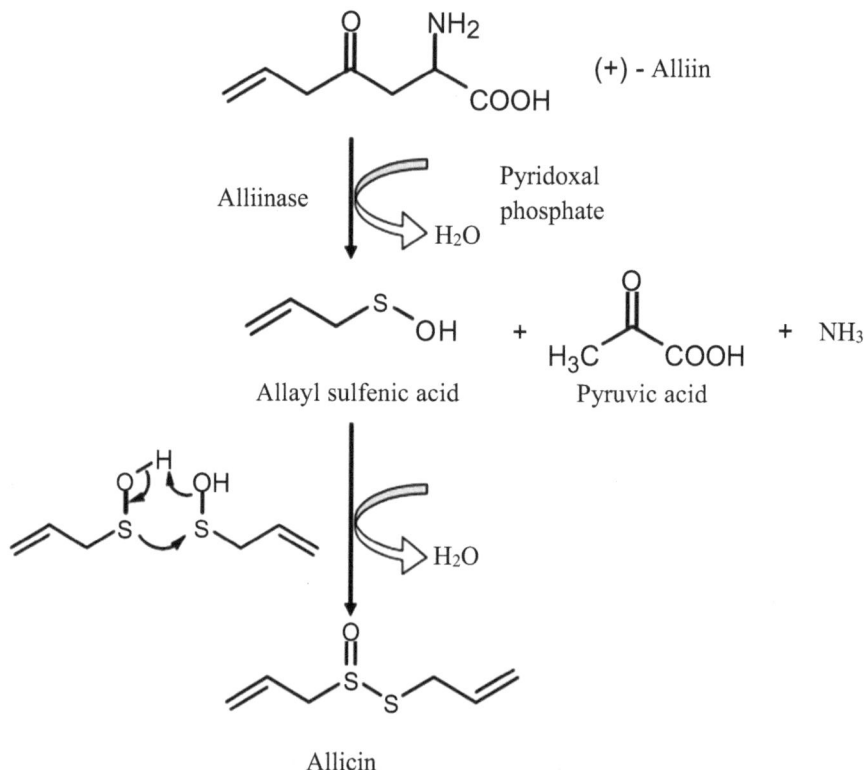

Fig. (1). Scheme of allicin biosynthesis.

These enzymatic transformations occur in 10-15 minutes at room temperature. The optimal conditions for alliinase activity are pH- 6.5 and temperature - 33°C. The activation energy of allicin decomposition is 14.7 kJ/mol [24]. Allicin is an oily liquid, bright yellow in color, with a characteristic garlic odor [25]. Allicin and allicin based products are highly unstable and volatile, this has made isolation and characterization of these products more problematic.

The chemical synthesis of allicin has become more relevant because of its commercial importance and the difficulty in isolation. The chemical synthesis is achieved through different methods, i) oxidation of allyl sulfide by hydrogen peroxide in acid medium [26, 27], ii) oxidation of allyl disulfide with m-chloroperbenzoic acid in chloroform [28], and iii) processing of dichloromethane solution of allyl disulfide by magnesium monoperoxy hydrate in the presence of ammonium-butyl sulfate [29]. Because of the high volatility and instability of allicin, all these chemical synthesis processes are typically carried out at low temperatures (0°C to room temperature), and depending on the method of synthesis and purification procedures, allicin of various purity levels are obtained. Allicin can disintegrate under the influence of various factors, temperature and pH are major among others, degradation occurs readily even at room temperature.

Degradation of allicin leads to generation of a series of predominant secondary products, mainly, (1) allyl methyl sulfide (AMS), (2) diallyl sulfide (DAS), (3) dimethyl disulfide (DMDS), (4) allylmethyl disulfide (AMDS), (5) dimethyl trisulfide (DMTS), (6) diallyl disulfide (DADS), (7) E/Z-4,5,9-trithiadode-a-1,6,11-triene 9-oxide, (8) *S*-allyl cysteine (SAC), (9) diallyltrisulfide (DATS), (10) 2-vinyl-2,4-dihydro-1,3-dithiin, (11) allyl methyl trisulfide (AMTS), (12) 3-vinyl-3,4-dihydro-1,2-dithiin, (*E*/*Z*-ajoene) and (13) *S*-allylmercaptocysteine (SAMC). Among these compounds 8 and 13 are water soluble compounds.

Medicinal Importance

Allicin demonstrates a extensive array of antibacterial properties towards several gram positive and gram negative bacteria, such as *Shigella* (*Shigella boydii, Shigella flexneri, Shigella sonnei*), *Salmonella enterica, Escherichia coli, Staphylococcus aureus, Enterococcus faecalis, Streptococcus (Streptococcus pyogenes, Streptococcus mutans, Streptococcus faecalis,), Klebsiella aerogenes Proteus vulgaris* and *Pseudomonas aeruginosa*. Fungi susceptible to allicin are *Aspergillus niger* and *Candida albicans* [30 - 35]. *In vitro* and *in vivo* antiviral activities of allicin were also reported. Major viruses sensitive to allicin are, *Parainfluenza virus type 3, Herpes simplex type 1* and *2, Vaccinia virus, Human Cytomegalo virus, Influenza B, Human rhinovirus type 2* and *Vesicular stomatitis virus* [36]. Interaction of allicin with vital thiol containing enzymes could be the major reason for its antimicrobial properties. A number of enzymes - alcohol dehydrogenases, cysteine proteinases and thioredoxin reductases - which are crucial for conserving the optimal redox state within amoeba and parasite were reported to be inhibited by allicin [37].

The antioxidant potential of many garlic fractions and formulations were established in different *in vitro* and *in vivo* studies (Fig. **2**). This antioxidant effect is linked primarily to the occurrence of allicin in garlic [30, 35, 38]. Studies have shown high antioxidant activity of allicin on the stable DPPH radical [30]. The capacity of garlic extracts and garlic organosulfur compounds to inhibit lipid peroxidation is linked to their free radical scavenging and antioxidant activities [39]. Studies have established the role of allicin in increased activity of anti-oxidant enzymes such as catalase [40], superoxide dismutase [41, 42] as well as increased glutathione content [40, 43, 44]. *In vivo* experiments have shown that allicin exhibits antioxidant effect in low concentrations, while a pro-oxidative effect in high concentrations [33].

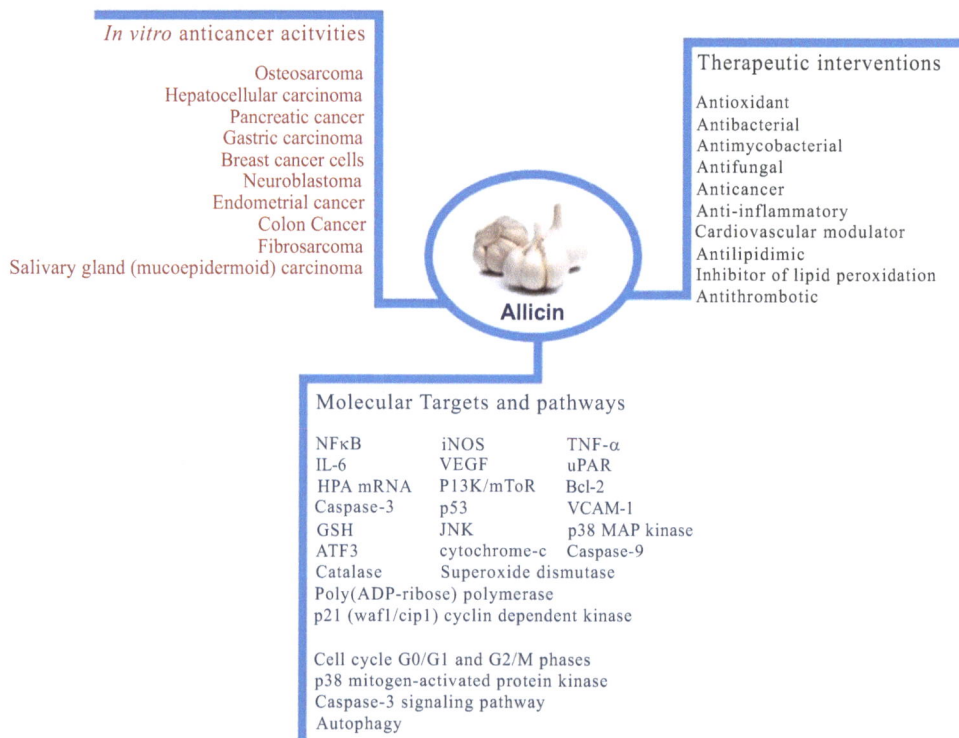

Fig. (2). Medicinal importance of allicin. A schematic diagram showing the reported therapeutic interventions, *in vitro* anticancer activities of allicin and the molecular targets and pathways of mechanism action of allicin.

Many biochemical effects of allicin has been reported including protective role in stroke [28], reduces the aggregation of thrombocytes [45, 46] reduces buildup of lipids, cholesterol, and calcium in large and medium arteries, hypertension and inhibits growth of cancerous cells [45, 47]. Negative impacts of allicin on thyroid function, growth, hyperplasia, allergic reactions on skin and mucous membrane were reported. It is reported that administration of allicin is hazardous to patients with disorders such as irregular blood pressure, diabetes, high cholesterol, patients with malignant diseases and gastrointestinal tract disorder [48]. Allicin is found to inhibit TNF-α and iNOS production, NFκB activation and IL6 [49, 50]. iNOS expression in activated macrophages was also inhibited by allicin [51].

Many recent studies have shown the apoptosis inducing and anticancer activities of allicin. Allicin inhibited proliferation of human colon (HT-29), mammary (MCF-7) and endometrial (Ishikawa) cancer cells (IC_{50}= 10-25 microM).

In mammary (MCF-7) cancer cells, increase of cells in the G0/G1 and G2/M phases of the cell cycle were observed with allicin induced growth inhibition. Allicin under *in vitro* conditions caused a momentary decrease in reduced glutathione, however the magnitude of which varied based on the cell type [52]. Allicin induced inhibition of metastasis of human colon carcinoma cell LoVo at non-cytotoxic concentration through down-regulating the expression of VEGF, uPAR and HPA mRNA [53]. Studies have shown that the levels of flavonoids, allicin and phenolic compounds have a great effect on the anticancer activities of garlic. This leads to the inference that fresh garlic possess high amount of biologically active molecules and the maximum anticancer property [54]. Allicin treatment resulted in decreased cell proliferation and high level expression of apoptosis related proteins in ATF3-overexpressed MCF-7 cells [55].

Studies on the *in vitro* and *in vivo* synergistic effect of antiproliferative compound artesunate with allicin in inhibiting the growth of human osteosarcoma cells, showed that the combination induces apoptosis by enhancing the activation of caspase - 3/9. This could be a promising combination in the treatment of osteosarcoma [56]. Allicin is reported to induce p53 mediated autophagy, and suppress the survivability of human hepatoma cells, which shows the declined p53 expression in the cytoplasm of HepG2 cells, inhibited PI3K/mTOR signal pathway, reduced Bcl-2 expression and enhanced signal transduction pathways of AMPK/TSC2 and Beclin 1 [57]. Similarly, in pancreatic cancer cells, allicin induced apoptosis through activation of caspase-3, cell cycle arrest, DNA fragmentation, ROS generation, p21(Waf1/Cip1) cyclin-dependent kinase inhibitor expression and GSH depletion [58]. Involvement of allicin in p38 mitogen-activated protein kinase/caspase-3 signaling pathway mediated apoptosis was established in human gastric carcinoma (MGC-803) cell lines [59]. Allicin inhibits the TNF-α-mediated induction of VCAM-1 through blocking ERK1/2 and NFκB signaling pathways and enhancing interaction between ER-α and p65, leading to the suppression of invasion and metastasis of MCF-7 cells [60].

In gastric cancer, allicin is reported to activate mitochondria-mediated apoptosis through endoplasmic reticulum (ER) stress and arresting cell cycle at G2/M phase. Apoptosis induction was mediated through both caspase - dependent/ independent and death receptor pathway [61]. Allicin activated the p38 MAPK pathway, leading to mitochondrial release of cytochrome-c, thus inducing apoptosis in human neuroblastoma SK-N-SH cells. The study suggests that allicin could be a novel agent in treatment strategies of neuroblastoma [62]. All these evidences lead to the conclusion that *Allium sativum* have many potential biomolecules that possess antimicrobial, antilipidemic, anticarcinogenic and antitumor properties. In summary, clinical and the epidemiological studies have proved that allicin could be the major biologically and pharmacologically active principle, which might be playing the crucial role in the biological activities of garlic.

CONCLUDING REMARKS

The available information clearly indicates the medicinal importance of allicin as molecule of great promise. The molecular targets and pathways evidently specifies the prominence of this molecule in various disease regulatory events. However, lack of *in vivo* studies makes it difficult to have a conclusive assessment. The lack of stability of the molecule is another problem which needs attention. Over all the molecule is a promising one with greater importance and needs more mechanistic studies for it's proper utilization.

CONFLICT OF INTEREST

The authors confirm that they have no conflict of interest to declare for this publication.

ACKNOWLEDGEMENTS

The authors thank Dr. PG Latha, Director (i/c), Kerala Forest Research Institute and Kerala State Council for Science, Technology and Environment (KSCSTE), Govt. of Kerala, India for providing necessary facilities and encouragement.

REFERENCES

[1]   Block E. The chemistry of garlic and onions. Sci Am 1985; 252(3): 114-9.
        [http://dx.doi.org/10.1038/scientificamerican0385-114] [PMID: 3975593]

[2]   Essman EJ. The medical uses of herbs. Fitoterapia 1984(55): 279.

[3]   Liu L, Yeh YY. Inhibition of cholesterol biosynthesis by organosulfur compounds derived from garlic.
        Lipids 2000; 35(2): 197-203.
        [http://dx.doi.org/10.1007/BF02664770] [PMID: 10757551]

[4]   Imai J, Ide N, Nagae S, Moriguchi T, Matsuura H, Itakura Y. Antioxidant and radical scavenging
        effects of aged garlic extract and its constituents. Planta Med 1994; 60(5): 417-20.
        [http://dx.doi.org/10.1055/s-2006-959522] [PMID: 7997468]

[5]   Rao RR, Rao SS, Natarajan S, Venkataraman PR. Inhibition of *Mycobacterium tuberculosis* by garlic
        extract. Nature 1946; 157: 441.
        [http://dx.doi.org/10.1038/157441b0] [PMID: 21066575]

[6]   Lau BH, Lam F, Wang CR. Effects of an odor-modified garlic preparation on blood lipids. Nutr Res
        1987; 7: 139-49.
        [http://dx.doi.org/10.1016/S0271-5317(87)80026-X]

[7]   Neil HA, Silagy CA, Lancaster T, *et al.* Garlic powder in the treatment of moderate hyperlipidaemia: a
        controlled trial and meta-analysis. J R Coll Physicians Lond 1996; 30(4): 329-34.
        [PMID: 8875379]

[8]   Hannan A, Ikram Ullah M, Usman M, Hussain S, Absar M, Javed K. Anti-mycobacterial activity of
        garlic (*Allium sativum*) against multi-drug resistant and non-multi-drug resistant *mycobacterium
        tuberculosis.* Pak J Pharm Sci 2011; 24(1): 81-5.
        [PMID: 21190924]

[9]   Herman-Antosiewicz A, Singh SV. Signal transduction pathways leading to cell cycle arrest and
        apoptosis induction in cancer cells by Allium vegetable-derived organosulfur compounds: a review.
        Mutat Res 2004; 555(1-2): 121-31.
        [http://dx.doi.org/10.1016/j.mrfmmm.2004.04.016] [PMID: 15476856]

[10]   Xiao D, Lew KL, Kim YA, *et al.* Diallyl trisulfide suppresses growth of PC-3 human prostate cancer
        xenograft *in vivo* in association with Bax and Bak induction. Clin Cancer Res 2006; 12(22): 6836-43. a
        [http://dx.doi.org/10.1158/1078-0432.CCR-06-1273] [PMID: 17121905]

[11]   Xiao XL, Peng J, Su Q, *et al.* [Diallyl trisulfide induces apoptosis of human gastric cancer cell line
        MGC803 through caspase-3 pathway]. Ai Zheng 2006; 25(10): 1247-51. b
        [PMID: 17059769]

[12]   Antosiewicz J, Herman-Antosiewicz A, Marynowski SW, Singh SV. c-Jun NH(2)-terminal kinase
        signaling axis regulates diallyl trisulfide-induced generation of reactive oxygen species and cell cycle
        arrest in human prostate cancer cells. Cancer Res 2006; 66(10): 5379-86.
        [http://dx.doi.org/10.1158/0008-5472.CAN-06-0356] [PMID: 16707465]

[13]   Block E, Ahmad S. (E, Z)-Ajoene a potent antithrombotic agent from garlic. J Am Chem Soc 1984;
        106: 95-6.
        [http://dx.doi.org/10.1021/ja00338a049]

[14]   Block E, Ahmad S, Catalfamo JL, Jain MK, Apitz CR. The chemistry of alkyl thiosulfinate esters. 9.
        Antithrombotic organosulfur compounds from garlic: structural, mechanistic and synthetic studies. J
        Am Chem Soc 1986; 108(22): 7045-55.
        [http://dx.doi.org/10.1021/ja00282a033]

[15] Schäfer G, Kaschula CH. The immunomodulation and anti-inflammatory effects of garlic organosulfur compounds in cancer chemoprevention. Anticancer Agents Med Chem 2014; 14(2): 233-40.
[http://dx.doi.org/10.2174/18715206113136660370] [PMID: 24237225]

[16] Cavallito CJ, Bailey JH. Allicin, the Antibacterial Principle of *Allium sativum*. I. isolation, physical properties and antibacterial action. Am Chem Soc 1944; 66(11): 1950.
[http://dx.doi.org/10.1021/ja01239a048]

[17] Zhang W, Ha M, Gong Y, Xu Y, Dong N, Yuan Y. Allicin induces apoptosis in gastric cancer cells through activation of both extrinsic and intrinsic pathways. Oncol Rep 2010; 24(6): 1585-92.
[PMID: 21042755]

[18] Louis XL, Murphy R, Thandapilly SJ, Yu L, Netticadan T. Garlic extracts prevent oxidative stress, hypertrophy and apoptosis in cardiomyocytes: a role for nitric oxide and hydrogen sulfide. BMC Complement Altern Med 2012; 12: 140.
[http://dx.doi.org/10.1186/1472-6882-12-140] [PMID: 22931510]

[19] Wang Z, Liu Z, Cao Z, Li L. Allicin induces apoptosis in EL-4 cells *in vitro* by activation of expression of caspase-3 and -12 and up-regulation of the ratio of Bax/Bcl-2. Nat Prod Res 2012; 26(11): 1033-7.
[http://dx.doi.org/10.1080/14786419.2010.550894] [PMID: 21902562]

[20] Cha JH, Choi YJ, Cha SH, Choi CH, Cho WH. Allicin inhibits cell growth and induces apoptosis in U87MG human glioblastoma cells through an ERK-dependent pathway. Oncol Rep 2012; 28(1): 41-8.
[PMID: 22552443]

[21] Kourounakis PN, Rekka EA. Effect on active oxygen species of alliin and *Allium sativum* (garlic) powder. Res Commun Chem Pathol Pharmacol 1991; 74(2): 249-52.
[PMID: 1667340]

[22] Ilic D, Nikolic V, Nikolic L, Stankovic M, Stanojevic L. Cakic, Milorad Allicin and related compounds: Biosynthesis, synthesis and pharmacological activity. Facta Universitatis 2011; 9(1): 9-20.
[http://dx.doi.org/10.2298/FUPCT1101009I]

[23] Peter KV. Handbook of Herbs and Spices. Boca Raton, FL: CRC Press LLC 2000; pp. 193-207.

[24] Mütsch-Eckner M, Erdelmeier CA, Sticher O, Reuter HD. A novel amino acid glycoside and three amino acids from *Allium sativum*. J Nat Prod 1993; 56(6): 864-9.
[http://dx.doi.org/10.1021/np50096a009] [PMID: 8350088]

[25] Sticher O. Beurteilung von Knoblauchpraparaten. Dtsch Apoth Ztg 1991; 131: 403-13.

[26] Freeman F, Kodera Y. Garlic chemistry: stability of (S)-(2-propenyl) 2-propene-1-sulfinothioate (Allicin) in blood, solvents, and simulated physiological fluids. J Agric Food Chem 1995; 43(9): 2332-8.
[http://dx.doi.org/10.1021/jf00057a004]

[27] Nikolić V, Stanković M, Nikolić Lj, Cvetković D. Mechanism and kinetics of synthesis of allicin. Pharmazie 2004; 59(1): 10-4.
[PMID: 14964414]

[28] Block E, Ahmad S, Catalfamo JL, Jain MK, Apitz CR. Antithrombotic organosulfur compounds from garlic: structural, mechanistic and synthetic studies. J Am Chem Soc 1986; 108: 7045-55.
[http://dx.doi.org/10.1021/ja00282a033]

[29] Cruz-Villalon G. Synthesis of allicin and purification by solid-phase extraction. Anal Biochem 2001; 290(2): 376-8.
[http://dx.doi.org/10.1006/abio.2001.4990] [PMID: 11237342]

[30]   Ilić DP, Nikolić VD, Nikolić LjB, Stanković MZ, Stanojević LjP. Thermal degradation, antioxidant and antimicrobial activity of the synthesized allicin and allicin incorporated in gel. Hemijskaindustrija 2010; 6(2): 85-93.

[31]   Nikolić V, Stanković M, Kapor A, Nikolić Lj, Cvetković D, Stamenković J. Allylthiosulfinate: beta-cyclodextrin inclusion complex: preparation, characterization and microbiological activity. Pharmazie 2004; 59(11): 845-8.
[PMID: 15587584]

[32]   Amagase H, Petesch BL, Matsuura H, Kasuga S, Itakura Y. Intake of garlic and its bioactive components. J Nutr 2001; 131(3s): 955S-62S.
[PMID: 11238796]

[33]   Ross ZM, OGara EA, Hill DJ, Sleightholme HV, Maslin DJ. Antimicrobial properties of garlic oil against human enteric bacteria: evaluation of methodologies and comparisons with garlic oil sulfides and garlic powder. Appl Environ Microbiol 2001; 67(1): 475-80.
[http://dx.doi.org/10.1128/AEM.67.1.475-480.2001] [PMID: 11133485]

[34]   Chen YY, Chiu HC, Wang YB. Effects of garlic extract on acid production and growth of Streptococcus mutans. J Food Drug Anal 2009; 17(1): 59-63.

[35]   Coppi A, Cabinian M, Mirelman D, Sinnis P. Antimalarial activity of allicin, a biologically active compound from garlic cloves. Antimicrob Agents Chemother 2006; 50(5): 1731-7.
[http://dx.doi.org/10.1128/AAC.50.5.1731-1737.2006] [PMID: 16641443]

[36]   Ankri S, Mirelman D. Antimicrobial properties of allicin from garlic. Microbes Infect 1999; 1(2): 125-9.
[http://dx.doi.org/10.1016/S1286-4579(99)80003-3] [PMID: 10594976]

[37]   Ankri S, Miron T, Rabinkov A, Wilchek M, Mirelman D. Allicin from garlic strongly inhibits cysteine proteinases and cytopathic effects of *Entamoeba histolytica.* Antimicrob Agents Chemother 1997; 41(10): 2286-8.
[PMID: 9333064]

[38]   Kim SM, Kubota K, Kobayashi A. Antioxidative activity of sulfur-containing flavor compounds in garlic. Biosci Biotechnol Biochem 1997; 61(9): 1482-5.
[http://dx.doi.org/10.1271/bbb.61.1482]

[39]   Rose P, Whiteman M, Moore PK, Zhu YZ. Bioactive S-alk(en)yl cysteine sulfoxide metabolites in the genus Allium: the chemistry of potential therapeutic agents. Nat Prod Rep 2005; 22(3): 351-68.
[http://dx.doi.org/10.1039/b417639c] [PMID: 16010345]

[40]   Sundaresan S, Subramanian P. Prevention of N-nitrosodiethylamine-induced hepatocarcinogenesis by S-allylcysteine. Mol Cell Biochem 2008; 310(1-2): 209-14.
[http://dx.doi.org/10.1007/s11010-007-9682-4] [PMID: 18185914]

[41]   Gudi VA, Singh SV. Effect of diallyl sulfide, a naturally occurring anti-carcinogen, on glutathione-dependent detoxification enzymes of female CD-1 mouse tissues. Biochem Pharmacol 1991; 42(6): 1261-5.
[http://dx.doi.org/10.1016/0006-2952(91)90263-5] [PMID: 1888335]

[42]   Perchellet JP, Perchellet EM, Abney NL, Zirnstein JA, Belman S. Effects of garlic and onion oils on glutathione peroxidase activity, the ratio of reduced/oxidized glutathione and ornithine decarboxylase induction in isolated mouse epidermal cells treated with tumor promoters. Cancer Biochem Biophys 1986; 8(4): 299-312.
[PMID: 3802049]

[43]   Ameen M, Musthapa MS, Abidi P, Ahmad I, Rahman Q. Garlic attenuates chrysotile-mediated pulmonary toxicity in rats by altering the phase I and phase II drug metabolizing enzyme system. J Biochem Mol Toxicol 2003; 17(6): 366-71.
[http://dx.doi.org/10.1002/jbt.10100] [PMID: 14708092]

[44]   Pinto JT, Qiao C, Xing J, *et al.* Effects of garlic thioallyl derivatives on growth, glutathione concentration, and polyamine formation of human prostate carcinoma cells in culture. Am J Clin Nutr 1997; 66(2): 398-405.
[PMID: 9250120]

[45]   Lawson LD. Garlic: a review of its medicinal effects and indicated active compounds. ACS Symposium Series. ACS Publications 1998; 691: pp. 176-209.
[http://dx.doi.org/10.1021/bk-1998-0691.ch014]

[46]   Fujisawa H, Suma K, Origuchi K, Kumagai H, Seki T, Ariga T. Biological and chemical stability of garlic-derived allicin. J Agric Food Chem 2008; 56(11): 4229-35.
[http://dx.doi.org/10.1021/jf8000907] [PMID: 18489116]

[47]   Miller KL, Liebowitz RS, Newby LK. Complementary and alternative medicine in cardiovascular disease: a review of biologically based approaches. Am Heart J 2004; 147(3): 401-11.
[http://dx.doi.org/10.1016/j.ahj.2003.10.021] [PMID: 14999187]

[48]   Koch HP. Analytische Bewertung von Knoblaucholmazeraten. Dtsch Apoth Ztg 1992; 27: 1419-26.

[49]   Bruck R, Aeed H, Brazovsky E, Noor T, Hershkoviz R. Allicin, the active component of garlic, prevents immune-mediated, concanavalin A-induced hepatic injury in mice. Liver Int 2005; 25(3): 613-21.
[http://dx.doi.org/10.1111/j.1478-3231.2005.01050.x] [PMID: 15910499]

[50]   Salman H, Bergman M, Bessler H, Punsky I, Djaldetti M. Effect of a garlic derivative (alliin) on peripheral blood cell immune responses. Int J Immunopharmacol 1999; 21(9): 589-97.
[http://dx.doi.org/10.1016/S0192-0561(99)00038-7] [PMID: 10501628]

[51]   Dirsch VM, Kiemer AK, Wagner H, Vollmar AM. Effect of allicin and ajoene, two compounds of garlic, on inducible nitric oxide synthase. Atherosclerosis 1998; 139(2): 333-9.
[http://dx.doi.org/10.1016/S0021-9150(98)00094-X] [PMID: 9712340]

[52]   Hirsch K, Danilenko M, Giat J, *et al.* Effect of purified allicin, the major ingredient of freshly crushed garlic, on cancer cell proliferation. Nutr Cancer 2000; 38(2): 245-54.
[http://dx.doi.org/10.1207/S15327914NC382_14] [PMID: 11525603]

[53]   Gao Y, Liu YQ, Cao WK, *et al.* Effects of allicin on invasion and metastasis of colon cancer LoVo cell line *in vitro.* Zhonghua Yi XueZaZhi 2009; 89(20): 1382-6.
[PMID: 19671326]

[54]   Shirzad H, Taji F, Rafieian-Kopaei M. Correlation between antioxidant activity of garlic extracts and WEHI-164 fibrosarcoma tumor growth in BALB/c mice. J Med Food 2011; 14(9): 969-74.
[http://dx.doi.org/10.1089/jmf.2011.1594] [PMID: 21812650]

[55]   Park B, Kyoungho K, Dong-kwon R, Suhkneung P. The apoptotic effect of allicin in MCF-7 human breast cancer cells: role for ATF3. FASEB J 2012; 26: lb367.

[56]   Jiang W, Huang Y, Wang JP, Yu XY, Zhang LY. The synergistic anticancer effect of artesunate combined with allicin in osteosarcoma cell line *in vitro* and *in vivo.* Asian Pac J Cancer Prev 2013; 14(8): 4615-9.
[http://dx.doi.org/10.7314/APJCP.2013.14.8.4615] [PMID: 24083713]

[57]   Chu YL, Ho CT, Chung JG, Rajasekaran R, Sheen LY. Allicin induces p53-mediated autophagy in Hep G2 human liver cancer cells. J Agric Food Chem 2012; 60(34): 8363-71.
[http://dx.doi.org/10.1021/jf301298y] [PMID: 22860996]

[58]   Chhabria SV, Akbarsha MA, Li AP, Kharkar PS, Desai KB. *In situ* allicin generation using targeted alliinase delivery for inhibition of MIA PaCa-2 cells *via* epigenetic changes, oxidative stress and cyclin-dependent kinase inhibitor (CDKI) expression. Apoptosis 2015; 20(10): 1388-409.
[http://dx.doi.org/10.1007/s10495-015-1159-4] [PMID: 26286853]

[59]   Zhang X, Zhu Y, Duan W, Feng C, He X. Allicin induces apoptosis of the MGC-803 human gastric carcinoma cell line through the p38 mitogen-activated protein kinase/caspase-3 signaling pathway. Mol Med Rep 2015; 11(4): 2755-60.
[PMID: 25523417]

[60]   Leea CG, Hee WL, Byung OK, Dong KR, Suhkneung P. Allicin inhibits invasion and migration of breast cancer cells through the suppression of VCAM-1: regulation of association between p65 and ER-α. J Funct Foods 2015; (15): 172-85.
[http://dx.doi.org/10.1016/j.jff.2015.03.017]

[61]   Luo R, Fang D, Hang H, Tang Z. Recent progress of allicin on cell growth inhibition and apoptosis in gastric cancer cells. Anticancer Agents Med Chem 2015; 16: 999.
[http://dx.doi.org/10.2174/1871520616666151111115443]

[62]   Zhuang J, Li Y, Chi Y. Role of p38 MAPK activation and mitochondrial cytochrome-c release in allicin-induced apoptosis in SK-N-SH cells. Anticancer Drugs 2016; 27(4): 312-7.
[http://dx.doi.org/10.1097/CAD.0000000000000340] [PMID: 26771864]

CHAPTER 3

Boswellic Acids as Potential Cancer Therapeutics

Manjeet Kumar[1], Arvind Kumar[1], Omkar P. Dhamale[2] and Bhahwal Ali Shah[1,*]

[1] *Natural Product Microbes, CSIR-Indian Institute of Integrative Medicine (CSIR), Canal Road, Jammu Tawi, India*

[2] *Sanofi Pasteur, 1 Discovery Drive, Swiftwater, Pennsylvania, USA*

Abstract: Cancer is the second leading cause of deaths worldwide, while it finds the top spot in diseases which still are not 100% curable. In the past few decades, a great deal of progress has been made in discovering new chemical entities, which enables us to understand the cause of cancer at cellular and molecular levels. In this regard, one of the naturally occurring triterpenoid class of compounds known as boswellic acids (BAs), have shown great potential for the development of new anticancer drugs. The interest in these type of triterpenoids has augmented since molecules such as NVX-207 and CDDO-Me have reached clinical trials. The alcoholic extract of the gum has also undergone clinical trials for the treatment of endotoxin induced hepatitis. Recently, the use of boswellic acid as well as its semi synthetic derivatives to treat cancer had been considered as an emerging concept in oncology as these have garnered considerable attention as a chemo-preventive and therapeutic agent in cancer.

Keywords: Anti-cancer, *Boswellia* sp., Boswellic acids, Pentacyclic triterpenes.

INTRODUCTION

Boswellic acids (BAs), pentacyclic triterpenoid class of natural products are widely known for their anti-inflammatory and anti-arthritic activities [1 - 3]. They inhibit 5-lipoxygenase, an enzyme that produces leukotriene, which is mainly responsible for the body inflammation and interfere with many other biological pathways. These complex scaffolds are generally available from natural sources and because of the numerous stereogenic centers in the aliphatic cyclic systems, their total synthesis and derivatization is relatively more challenging. This class of compounds has provided promising leads for the development of new anti-cancer drugs [4]. The interest for these types of pentacyclic triterpenoids has also grown

---

* **Corresponding author Bhahwal Ali Shah:** Natural Product Microbes, CSIR-Indian Institute of Integrative Medicine (CSIR), Canal Road, Jammu Tawi, India; Tel/Fax: +91-191-2585006-10; EPABX: 311; E-mail: bashah@iiim.ac.in

Sahdeo Prasad & Amit Kumar Tyagi (Eds.)

since the clinical trials for NVX-207 (**1**, GZ 68·205/53-BrGT/2007) and CDDO-Me (**2**, https://clinicaltrials.gov/ct2/results?term=CDDO+methyl+ester&Search=Search) for cancer treatment (Fig. **1**) [5, 6].

Fig. (1). Triterpenoids under clinical evaluation.

In the recent past, several reviews have been published, highlighting the chemistry and biology of BAs, which can be consulted for more details in their respective areas [1 - 3, 7 - 11]. This chapter serves as a summary of isolation, characterization, and biological scope focused on anti-cancer studies and recent structure-activity relationships of newly developed BAs analogues.

History and Background

BAs are the important constituents of *Boswellia* genus, which contain almost 25 species, widely distributed in the dry areas of the Horn of Africa, the Arabian Peninsula and in India (Table **1**) [12 - 19]. Of all, *B. Serrata* is one of the most attractive and highly investigated species, generally found in subcontinent of India (Western Himalayas, dry hill forests of Rajasthan, Gujarat, Maharashtra, Madhya Pradesh, Bihar, Orissa). It belongs to the Burseraceae family, commonly known as frankincense, shallaki, salai guggal, white guggal, Indian olibanum or dhup, having a long history of use as incense in religious and cultural ceremonies [20]. In addition to this, there are written evidences where frankincense had been documented as drug in various diseases [21 - 34]. Probably, the oldest pharmacological note, the papyrus Ebers (received by a Professor of Egyptology, Moritz Fritz Ebers in 1873 from an Arabian businessman, found in between the legs of a mummy of Luxor), quoted about 900 medical prescription regarding the practical information of diagnosis and treatment of internal diseases, had also mentioned frankincense as a drug used for treatment of various diseases. The age of the Papyrus Ebers is dated to the time of Pharaoh Amenophis I and was most likely written around 1500 BC [35]. The use of the oleogum resin of *B. serrata* is also described in Ayurvedic text books (Charaka Samhita, 1st 2nd century AD and in Astangahrdaya Samhita, 7th century AD). Medical preparations containing the bark or the oleogum resin of *B. serrata* were used to treat a variety of diseases including tumors, carcinomas and oedemas. Moreover, respiratory tract like cough, other respiratory problems as well as diarrhoea, constipation, flatulence and central nervous diseases were also treated.

Table 1. Some of the important *Boswellia* species and their geographical distribution.

Species	Geographical Distribution
B. ovalifoliolata Bal. & Henry	India
B. pirottae Chiov.	Ethopia
B. neglecta S. Moore	Somalia, Kenya
B. dalzielii Hutch.	Tropical Africa
B. papyrifera Hochst.	Ethiopia, Eritrea, Sudan
B. rivae Engl.	Ethopia
B. hildebrandtii Engl.	Somalia
B. ogadensis Vollesen	Ethopia
B. popoviana Hepper	Yemen
B. nana Hepper	Yemen
B. dioscorides Thul. & Gifri	Yemen
B. bullata Thul. & Gifri	Yemen
Boswellia carterii Birdw.	Somali, Nubia
Boswellia sacra Flück.	Oman, Yemen
Boswellia frereana Birdw.	Somalia
Boswellia odorata Hutch.	Tropical Africa
Boswellia ameero Balf. Fils.	Yemen, Socotra
Boswellia elongata Balf. Fils.	Yemen, Socotra
Boswellia socotrana Balf. Fils.	Socotra
Boswellia serrata Roxb.	India

Botanical and Geographical Distribution

Boswellia resins are harvested during the start of hotter period of the year. After a certain time the deciduous tree exude viscous liquid, which on air exposure transforms into a solid gum resin. Depending upon the age, height and condition of tree, its exploitation is executed for three consecutive years and yields about 3-10 kg of oleo-gum resin [36]. More than 200 compounds have been isolated from the oleogum resin of different Boswellia species including polysaccharides, essential oils, proteins and inorganic compounds [37 - 40]. However, the major constituents of oleogum resin are essential oils, polysaccharides and higher terpenoids, and their percentage differs from species to species as well as depends upon the harvesting conditions and locations. Essential oils generally consist of volatile compounds like terpenoids including monoterpenes, diterpenes, triterpenes, sesquiterpenes and low molecular weight aromatic compounds, present in the various parts of the plants [41, 42]. They are generally obtained from the low boiling fractions and the most common technique used for their isolation is hydro-distillation. Essential oils also find application in perfume and cosmetic industries. The other major part of oleogum resin consists of non-volatile compounds, which are generally excreted from the plant at the end of the metabolic pathways. Mostly these contains higher terpenoids such as tetracyclic triterpeniods *e.g.* 3-α-oxotirucallic acid, 3-α-hydroxytirucallic acid, 3- α-acetoxytirucallic acid and 3-β-acetoxytirucallic acid; pentacyclic triterpenoids *e.g.* α/β-boswellic acid, acteyl-α/β-boswellic acid, 11-keto-β-boswellic acid, acetyl-1--keto-β-boswellic acid, α-boswellic acid, acetyl-α-boswellic acid *etc.* Among these, boswellic acids are mainly responsible for the pharmacological effect of oleogum resin. In the recent past, a significant decline in the overall production of frankincense has been observed in different countries with the exception in India. This can be attributed to various reasons; in Oman, oil export is the major source of economy involving maximum of the workforce coupled with harsh conditions like heat and dry climate making harvesting difficult [43]. In Yemen, problem is the availability of place, where lot of field space is being used for producing drug kath (*Catha edulis*) [44]. Conditions are not good in Africa, Ethiopia, Eritrea and Djibouti possibly because of economy crisis, terrorism and border conflicts [45]. With the exception, India have seen a significant growth in the production of *B. Serrata,* which seem to be because of the lot of research being carried out on the extracts of *B. serrata* over the last two decades [46]. This may also be due to the BAs, which are the major constituents of *B. Serrata* as well as responsible for the pharmacological applications such as anti-inflammatory [47 - 53], anti-arthritic [54, 55], anti-cancer [56 - 63], anti-oxidant [64, 65], anti-ulcer [66, 67], anti-bacterial [68 - 70], anti-asthmatic [71], anti-atherosclerotic [72], anti-diarrheal [73], hepatoprotective [74], renoprotective [75], anti-hyperglycemic [76, 77], wound healing [78], diuretic [79], analgesic [80] and also the ingredient of certain ointments for rheumatism [81] and nervous disorders [82].

Boswellic Acids:

Although investigation of different constituents from oleogum resin has a long history, but it was not until the last of 18th century, when the first authentic report was published by Tschirch and Halbey [83], isolated an acidic constituent from the olibanum, known as Boswellic acid and assigned the molecular formula $C_{32}H_{52}O_x$. Although the structure of the compound was not known at that time, but efforts continued, and identification of isoprene rule by Wallack in same decade further fueled up the investigation. As a result, in 1932, Winterstein and Stein [84], first time isolated the pure four compounds from the same resin and shown that Tschirch and Halbey had actually isolated the mixture of four boswellic acids, namely the α-boswellic acid (**3**, partially characterized), β-boswellic acid (**4**, partially characterized) and their acetylated derivatives (**5,6**, partially characterized) (Fig. **2**). They modified the chemical formula to $C_{30}H_{48}O_3$ and characterized the molecules as monobasic monohydroxy acid, but again the chemical structures were not completely characterized.

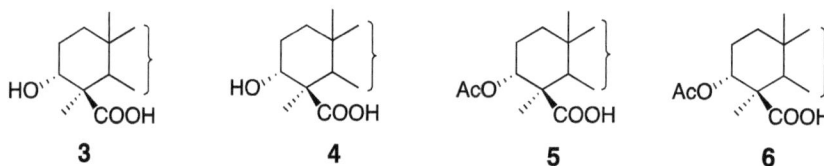

Fig. (2). Partially characterized structures of triterpenes.

In the same decade, a lot of results were published regarding the structural characterization of boswellic acid, including the major contribution by Simpson *et al.* [85, 86] and Ruchika *et al.* [87 - 91] independently, in describing the basic skeleton. Simpson *et al.* [85] reported the oxidation of β-boswellic acid, thus confirming the presence of secondary hydroxy group in the system. The oxidized molecule was characterized as nor-β-boswellenone (**7**) having molecular formula $C_{29}H_{46}O$ (Fig. **3**). Further, oxidation of the molecule in the presence of $KMnO_4$, resulted in the formation of another ketone functionality, but remained unassigned at that time. In the same year, same group showed the presence of double bond functionality in the molecule, which was confirmed by the allylic oxidation of methylene group in the presence of chromic acid [86]. The molecule further assigned the name as nor-β-boswellanedione (**8**), having molecular formula $C_{29}H_{46}O_2$ (Fig. **3**). Ruchika and coworkers also confirmed the results, where they investigated the core structure by chemical transformation of β-boswellic acid into different triterpenic compounds [87 - 91]. This partial information of the various functionalities and the resemblance of a basic core structure with known triterpenes had attracted interests of many research groups at that time. This can be easily visualized from lot of isolation and characterization publications reported in the same decade [92 - 95].

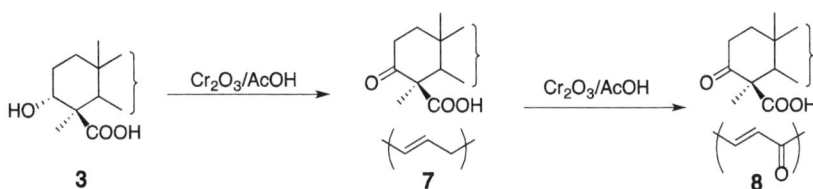

Fig. (3). Partial structure characterization of triterpenes.

In 1956, Beton *et al.* [96], first time reported the complete structural information of boswellic acid as an ursane analogue having 3α-axial configuration of the OH group and the 4β-axial configuration of the CO_2H group *i.e.* 3α-hydroxyurs-12-en-24-oic acid. In the same decade, a major breakthrough contribution in the confirmation of molecular structures came from Shamma and coworkers [97], where they reported the NMR spectroscopic data of the range of pentacyclic triterpenic compounds including boswellic esters. However, it took almost two decade to publish the complete characterization data of BAs derivatives, when Pardhy and Bhattacharyya [98], in 1978, reported the isolation of four BAs *i.e.* β-boswellic acid (**9**), acetyl-β-boswellic acid (**10**), 11-keto-β-boswellic acid (**11**) and acetyl-11-keto-β-boswellic acid (**12**) from the resin of *B. serrata Roxb* and first time presented the mass, IR and 1H NMR spectroscopic data of compounds (Fig. **4**).

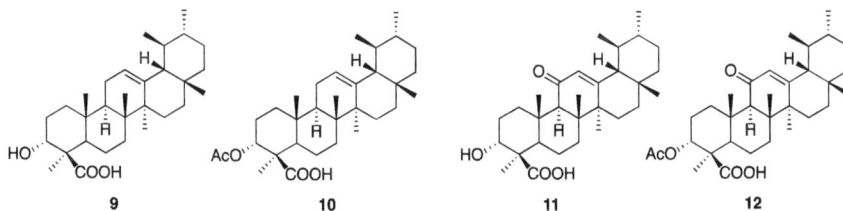

Fig. (4). Structures of boswellic acids.

The successful interpretation of these compounds resulted in the isolation and characterization of many other pentacyclic triterpenes by the end of the last century.

So far fourteen pentacyclic triterpenes have been isolated from *B. serrata* including α/β-boswellic acid (**13,9**), O-acetyl-α/β-boswellic acid (**14,10**), 11-ket--β-boswellic acid (**11**), O-acetyl-11-keto-β-boswellic acid (**12**), O-acetyl-11-OH-β-boswellic acid (**15**), 11-OMe-O-acetyl-β-boswellic acid (**16**), 9,11-dehydr--α/β-boswellic acid (**17,18**), 9,11-dehydro-O-acetyl-α/β-boswellic acid (**19,20**), leupolic acid (**21**), acetyl leupolic acid (**22**) (Fig. **5**).

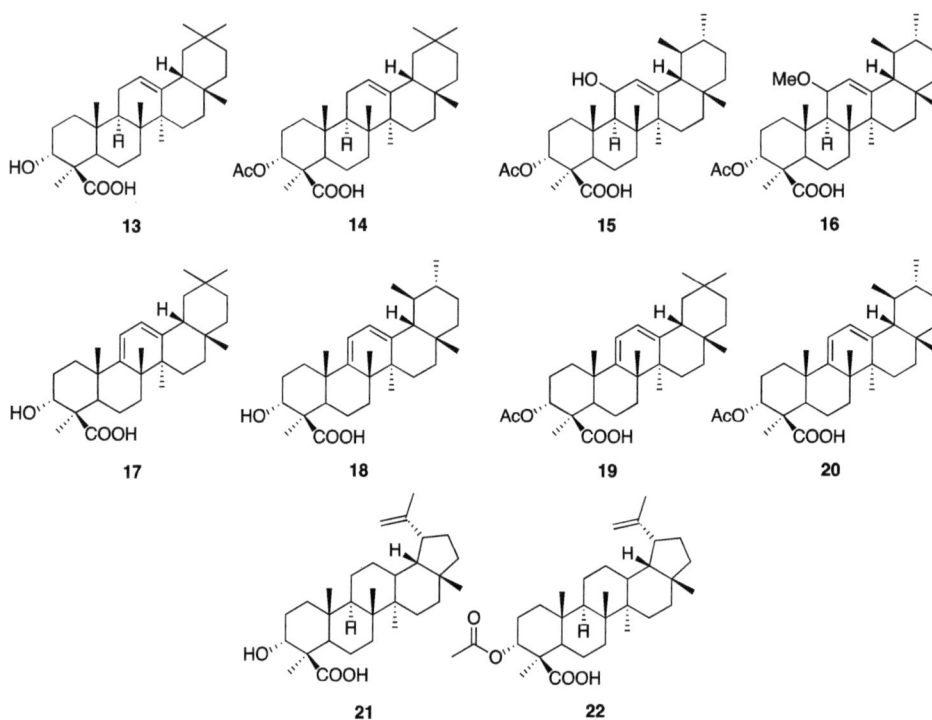

Fig. (5). Pentacyclic triterpenes isolated from *B. serrata*.

Biological Studies

Frankincense resins have extensively been used by ancient civilizations of the Indians and Africans for the treatment of various diseases. At present, there is a great deal known regarding the therapeutic usage of *B. species* in anti-inflammation and anti-arthritis. Several reviews have covered the medicinal aspect of *B. species* in these areas [1 - 3, 5, 6]. Since as the pure molecules have been isolated from various *B. species*, new therapeutic potential of these resins have been explored. In this chapter, we focus on the anti-proliferative potential of BAs *via* different mechanistic pathways and tried to analyze the biological impact of substitution at each of the accessible carbon in pentacyclic triterpene scaffold.

Topoisomerase Inhibitors

The very first evidence regarding the anti-cancer potential of BAs came form Chinese clinical studies, where they reported the anti-tumor activity of the Chinese herbal medicine Tian-shian-Wan. Since, the chemical constituents and their composition was not present at that time, so again the attempts were made by Cook *et al.*, [99] in 1990, where the sample of that medicine was subjected to the chromatographic fractionation. The bio-guided fraction resulted in identification of two compounds, which are mainly responsible for bioactivity of the extract. These compound were isolated through HPLC and characterized by mass spectral analysis, as $C_{32}H_{50}O_4$, found to be the acetates of α/β-boswellic acids (**14,10**). The compounds were confirmed by 1D, 2D, NMR spectroscopy and X-ray crystallography. Both compounds were found to be topoisomerase I and II inhibitors, whereas α-boswellic acid shows better potency in the topoisomerase I inhibition assay. In addition, compounds were also found to induce cell differentiation in HL-60 cells lines at 10 μg/ml. *In vivo* results further confirmed the anticancer potency of compounds, where four out of ten tumor bearing mice survived when subjected to testing compounds (1:1 mixture of both compounds), while all of the mice in the control group died. The importance of boswellic acid acetates in topoisomerase inhibition is further exemplified by Hoernlein *et al.* [100] who showed that acetyl-11-keto-β-boswellic acid (**12**) induces apoptosis in HL-60 and CCRF-CEM cell lines (IC50 = 30 μm). However, 11-keto-BAs are well known 5-lipoxygenase inhibitors, which may possibly trigger apoptosis, but since HL-60 and CCRF-CEM cell lines do not express 5-lipoxygenase mRNA essentially, so the authors reasoned that there would be some other mechanistic pathway responsible for apoptosis. They investigated the topoisomerases expression in both the cell lines and found that there is a significant reduction in expression of three topomerisaes *i.e.* topoisomerase I, topoisomerase II, and topoisomerase II mRNA, when treated with **12**, which may be the possible reason for induction of apoptosis. The cytotoxic effects of acetyl BAs were also reported by Shao *et al.* [101, 102], where the compounds were found to inhibit the DNA synthesis and reduce the cell proliferation in dose dependent manner. The rate of DNA, RNA and protein synthesis was calculated by rate of incorporation of thymidine, uridine and leucine into trichloroacetic acid insoluble material. The 50% inhibitory activity of four compounds in DNA synthesis was reported: thymidine/uridine/leucine: 3.7/7.1/6.3, 1.4/2.3/5.4, 0.9/2.2/5.1, 0.6/0.5/4.1 respec-tively. In the early 2000, Syrovets and coworkers [103] reported the mechanistic investigation of anti-proliferative activity of acetyl BAs. Studies indicate that immobilized acetyl-BAs binds directly to topoisomerases I and II through high-affinity binding sites yielding KD values of 70.6 nM for topoisomerase I and 7.6 nM for topoisomerase II, thus competing with DNA for binding to the enzyme. Additionally, the evidence also came from our lab [104, 105], where three 3-acyl derivatives *i.e.* 3-propionloxy-11-keto-β-boswellic acid (**23**), 3-butyryloxy--1-keto-β-boswellic acid (**25**) and 3-hexanoyloxy-11-keto-β-boswellic acid (**27**) have been reported to induce cytotoxicity in various cancer cell lines. All the molecules were found to induce apoptosis *via* inhibiting enzyme activity of topoisomerase I and II, while **25** was found to be the best topoisomerase inhibitor at 10 μg/ml (Fig. **6**).

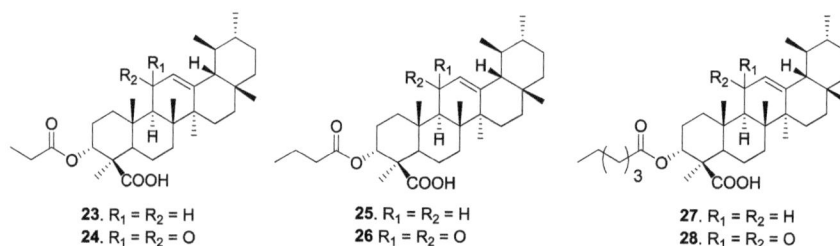

23. R₁ = R₂ = H   **25.** R₁ = R₂ = H   **27.** R₁ = R₂ = H
24. R₁ = R₂ = O   **26** R₁ = R₂ = O   **28.** R₁ = R₂ = O

Fig. (6). Acyl derivatives of boswellic acids.

NF-κB Signalling

In addition to the cell differentiation and anti-proliferation activity in leukemia cell lines, BAs, O-acetyl-β-boswellic acid (**10**) and O-acetyl-11-keto-β-boswellic acid (**12**) were also found to induce apoptosis in chemo-resistant androgen-independent PC-3 prostate cancer cells [106]. Mechanistically both molecules were found to interfere with the IκB kinase activity, which in turn inhibits the activated NF-κB signalling. Results were further confirmed by Yuan *et.al* [107]. and Pang *et.al.*, [108] where they also reported the anti-proliferative activity of **12**, in prostate cancer cells, although different pathways have been reported *i.e.* reduction in AR expression at mRNA and protein levels; angiogenesis suppression respectively. Since, triterpenes are well known for their anti-inflammatory properties and previous literature reports have revealed the complex roles of NF-κB signalling in inflammation, suppression of apoptosis, cell differentiation, growth promotion, angiogenesis and invasiveness. In a study that focused on NF-κB signalling for apoptosis induced by BAs, Takada *et al.* [109], reported that 3-*O*-acetyl-11-keto-β-boswellic acid (**12**) suppressed both inducible and constitutive NF-κB activation in tumor cells. In addition, it also diminished the effect of NF-κB activators like IL-1β, okadaic acid, doxorubicin, LPS, H_2O_2, PMA and cigarette smoke. At the molecular level, compound found to inhibit sequentially the TNF-induced activation of IκBα kinase (IKK) through AKT kinase inhibition, which in turn suppresses NF-κB gene expression. In order to confirm the potential of triterpenes in suppressing NF-κB activation, our group [110] also reported a novel semi-synthetic derivative, 3-O-butyryl-11-k-to-β-boswellic acid, which showed (**26**) significant cytotoxicity against a range of human cancer cell lines and mechanistically demonstrated an inhibitor of the NF-κB and STAT proteins. Recently, an additional evidence regarding the NF-κB inhibition by anti-inflammatory agents was published by Patil and coworker

[111], where they investigated a wide range of pentacyclic triterpenes (109 compounds) for anti-cancer activity and reported that most of the compounds were found the induce IKKβ inhibition, which is mainly responsible NF-κB inhibition.

Caspase Activation

Caspases are a family of endoproteases, present as zymogens (inactive form) in almost all healthy cells. Although, there are almost 11 caspases described in human, but not all are known to execute apoptosis. Among these, caspases 2, -8, -9 and -10 are known as initiator caspases, mainly play an important role in maintaining homeostasis through regulating cell death and inflammation. In a report by Liu *et al.* [112], BAs were found to interfere in caspase activation. They reported the anti-proliferative activity of 11-keto-β-boswellic acid (**11**) and O-acetyl-11-keto-β-boswellic acid (**12**), on liver cancer Hep G2 cells. Compounds were found to decrease the cell viability and inhibited the [3H] thymidine incorporation (50% and 15% of the control at 25 micro gram) in Hep G2 cells. At the molecular level, compounds were found to induce cell death through caspase activation. After this first report, several other groups have reported the role of BAs in caspase activation, which in turn triggers apoptosis *via* different pathways [113 - 115].

PI3/AKT Kinase Inhibitors

Liu and Duan [116] investigated the role of 3-*O*-acetyl-11-keto-β boswellic acid in PI3K kinase inhibition in the absence and presence of phosphatidylinositol-3-kinase (PI3K) inhibitor *i.e.* LY294002. Experimental results showed that AKBA induce apoptosis in the absence of LY294002, while addition of LY294002 to the cell culture significantly enhances apoptosis by 20 fold. Mechanistic studies revealed that AKBA induced phosphorylation of Akt at both Ser473 and Thr308 positions, indicating an activation of the PI3K/Akt pathway. Furthermore, we also reported a new derivative of 11-keto-β-boswellic acid *i.e.* butyl 2-cyano-3,11-dioxours-1,12- dien-24-oate (**29**), capable of inducing cell death in range of human cancer cell lines (IC$_{50}$; 0.67 mM/HL-60, 1 mM/Molt4 and 1.5 mM/THP1) [117]. At molecular level, compound was found to inhibit PI3K/Akt activity, NF-kB, Hsp-90, and surviving which may be responsible for apoptosis. Later, Morad *et al.* [118] identified of a novel semisynthetic triterpene derivative, 3-cinnamoyl-11-keto-β-boswellic acid **30** capable of inducing cell death *in vitro* cell lines as well as in PC-3 prostate cancer xenografts *in vivo via* mTOR inhibition. Mechanistic investigation showed that the compound binds to the FKBP12-rapamycin-binding domain of mTOR with high affinity, thereby competing with the endogenous mTOR activator phosphatidic acid. Also recently, Kumar and coworkers [119] also showed the strong inhibitory effect of 3-butyroloxy-11-k-to-β-boswellic acid (26) on PI3K/Akt or Ras/Raf/MEK/ERK pathways. In addition, BAs also induced cell death in different cell lines derived from malignant gliomas [120]. The cell death induced by boswellic acid was independent of reactive oxygen species, as various free radical scavengers could not attenuate the effect of boswellic acids induced apoptosis. However, chemical mediated inhibition of protein synthesis reversed the cell death induced by boswellic acid. BA induced p21 expression in different cell lines, which in turn inhibits the cyclin dependent kinases and thus induce apoptosis by augmenting cell cycle arrest at G0/G1 phase. This study also suggested that the cell death induced by boswellic acid was partially due to RNA synthesis inhibition (Fig. 7).

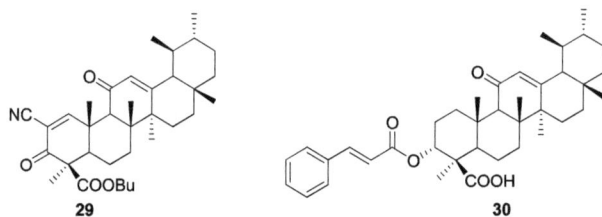

Fig. (7). Acyl derivatives of boswellic acids.

Other Targets

B. Serrata gum resin has also been utilized for the synthesis of BAs nanoparticles. Devika *et al.* [121] in 2013, reported the synthesis of milky white nanosuspension of boswellic acid nanoparticles having a particle size ranging from 150–190 nm. BA nanoparticles were investigated for the treatment of the tumor-bearing BALB/c mice. Experimental results showed the reduction in PC3 tumor-bearing BALB/c mice, when treated with boswellic acid oral nanoparticles (10 mg/kg, 20 mg/kg) for 14 days after maximal tumor growth size of about 0.525 ± 0.3535 unit with its corresponding tumor volume as 65.625 ± 2.2097 unit. In extension to this, recently Snima *et.al.*, [122] used boswellic acid nanoparticle in combination therapy. They reported the synergistic effect of BAs nanoparticles and anti-diabetic drug metformin in inhibition of cell proliferation in pancreatic cancer cell line. In addition, *B. serrata* have also been investigated in combination with doxorubicin for apoptosis in hepatocellular carcinoma cell lines. Studies showed that *B. serrata* act synergistically with doxorubicin against HepG2 and Hep3B cells and induce dose dependent increase in caspase-3 activity, TNF-α level and IL-6 level [123]. The synergistic effect of BAs and curcumin in anti-cancer studies have also been reported by Toden *et.al.*, [124] where curcumin and

O-acetyl-11-keto-β-boswellic acid synergistically found to induce apoptosis and cell-cycle arrest in colorectal cancer cell lines. At molecular level, both curcumin and boswellic acid induced up-regulation of tumor-suppressive miR-34a and down-regulate miR-27a in colorectal cancer cells. However later on, the same group performed the mechanistic investigation and suggested the different possible pathways for induced apoptosis. BAs were also reported as epigenetic modifiers, where the most active BAs, O-acetyl-keto-β-boswellic acid was shown to modulate the demethylation in various colorectal cancer cell lines [125, 126].

Structure Activity Relationship

In order to study the structure activity relationship of BAs, lot of chemistry has been evolved in the recent past. Presently, a lot of information is available regarding the effect of various substituents on different biological activities of the compounds. Here, we focused on the work that has led to the important understanding of the anti-cancer potential of compounds. Several reviews are available related to the other biological activates.

Generally, these complex scaffolds are available from natural sources because aliphatic cyclic systems having numerous stereogenic centers make their total synthesis and derivatization relatively challenging. Therefore, the synthesis of these complex scaffolds in the larger quantities is highly uneconomic, and semi-synthetic strategies represent the best alternatives to generate the library of molecules. The most studied area belongs to 3-hydroxy position of BAs. Since, the literature studies showed that 3-O-acetyl group *i.e.* in other words protection at 3-hydroxy position is very crucial in many biological activities, thus, the very much important structural modification involves the incorporation of acyl groups having different carbon chain lengths. In this regard, our group have reported the the synthesis of acyl derivatives of β-boswellic acids, 11-keto-β-boswellic acids and their epimers. The acyl analogues acids have been synthesized by treating **9** and **11** with respective anhydrides, in the presence of dimethylaminopyridine. In addition, epimerization of 3-hydroxy group of BAs have been done by first oxidation of 3-hydroxy group with pyridinium chlorochromate (PCC) and followed by reduction with $NaBH_4$ in methanol. Compounds were found to possess cytotoxicity against the range of human cancer cell lines. SAR indicated that acyl homologues increases the cytotoxicity of compounds, carboxylic acid group is important for activity as the esterification of acid group led to the reduction of activity, also the epimerization of 3-hydroxy group resulted in loss of activity **(37-44)**, which shows 3-α-hydroxy configuration is important for better cytotoxicity. The C-3 hemi-succinates of boswellic acids **(32, 34, 40, 44)** were also synthesized and screened for cytotoxicity. However, this modification resulted in the loss of cytotoxicity (Scheme **1**).

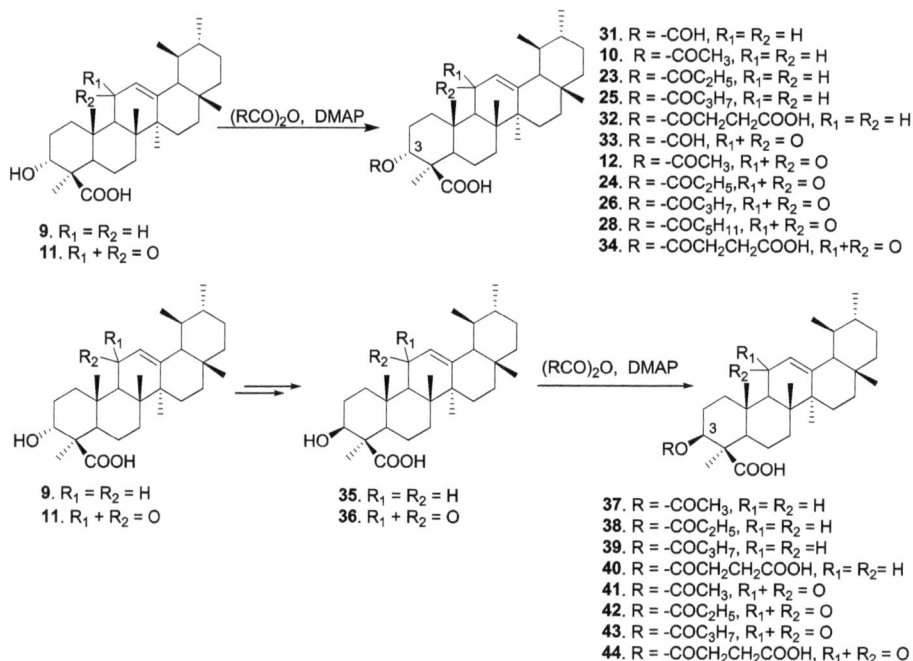

31. R = -COH, R_1= R_2 = H
10. R = -COCH$_3$, R_1= R_2 = H
23. R = -COC$_2$H$_5$, R_1= R_2 = H
25. R = -COC$_3$H$_7$, R_1= R_2 = H
32. R = -COCH$_2$CH$_2$COOH, R_1 = R_2 = H
33. R = -COH, R_1+ R_2 = O
12. R = -COCH$_3$, R_1+ R_2 = O
24. R = -COC$_2$H$_5$, R_1+ R_2 = O
26. R = -COC$_3$H$_7$, R_1+ R_2 = O
28. R = -COC$_5$H$_{11}$, R_1+ R_2 = O
34. R = -COCH$_2$CH$_2$COOH, R_1+R_2 = O

9. R_1 = R_2 = H
11. R_1 + R_2 = O

9. R_1 = R_2 = H
11. R_1 + R_2 = O

35. R_1 = R_2 = H
36. R_1 + R_2 = O

37. R = -COCH$_3$, R_1= R_2 = H
38. R = -COC$_2$H$_5$, R_1= R_2 = H
39. R = -COC$_3$H$_7$, R_1= R_2 =H
40. R = -COCH$_2$CH$_2$COOH, R_1= R_2= H
41. R = -COCH$_3$, R_1+ R_2 = O
42. R = -COC$_2$H$_5$, R_1+ R_2 = O
43. R = -COC$_3$H$_7$, R_1+ R_2 = O
44. R = -COCH$_2$CH$_2$COOH, R_1+ R_2 = O

Scheme 1. Synthesis of acyl analogues of boswellic acids.

The 4-amino analogues of β-boswellic acid and 11-keto-β-boswellic acids have also been investigated [127]. The curtius rearrangement of BAs followed by treatment with KOH resulted in the formation of amino alcohols (**45,46**). Different anhydride were employed in presence of DMAP to provide **47-52** (Scheme **2**). Using the same procedure, the amino analogues of *epi*-3-alkox--β-boswellic acid had also been synthesized **55-57**. All the semi-synthetic molecules were subjected to *in vitro* screening for anticancer activity against different human cancer cell lines. The results revealed that 4-α-amino analogues of boswellic acids exhibited better cytotoxicity than the corresponding carboxylic acid analogues. Also, the amino-acyl derivatives were found to exhibit lower activities as compare to amino-alcohols. In addition, the amino analogue of epi-BAs (**53,54**) displayed lower activity than amino BAs, which confirms the importance of α-configuration of alcohol at C-3 position.

Csuk and co-workers [128] synthesized an endoperoxide derivative of 11-keto-β-boswellic acid (Scheme **3**). The acid group of 11-keto-β-boswellic acid was subjected to esterification with methyl iodide and K$_2$CO$_3$, and the corresponding ester was treated with 3,3-dimethylglutarimide, triphenylphosphine and DEAD to afford the **58**. The diene analogue **58** was successfully transformed into endoperoxide **59** by treatment with acetic acid, N-hydroxy-succinimide and $Na_2Cr_2O_7$. The endoperoxide **59** displayed high antitumor activity against 15 human cancer cell lines, having minimum IC50 of 0.4 μM against anaplastic thyroid and induced apoptosis evidenced by a dye-exclusion test and DNA laddering experiments.

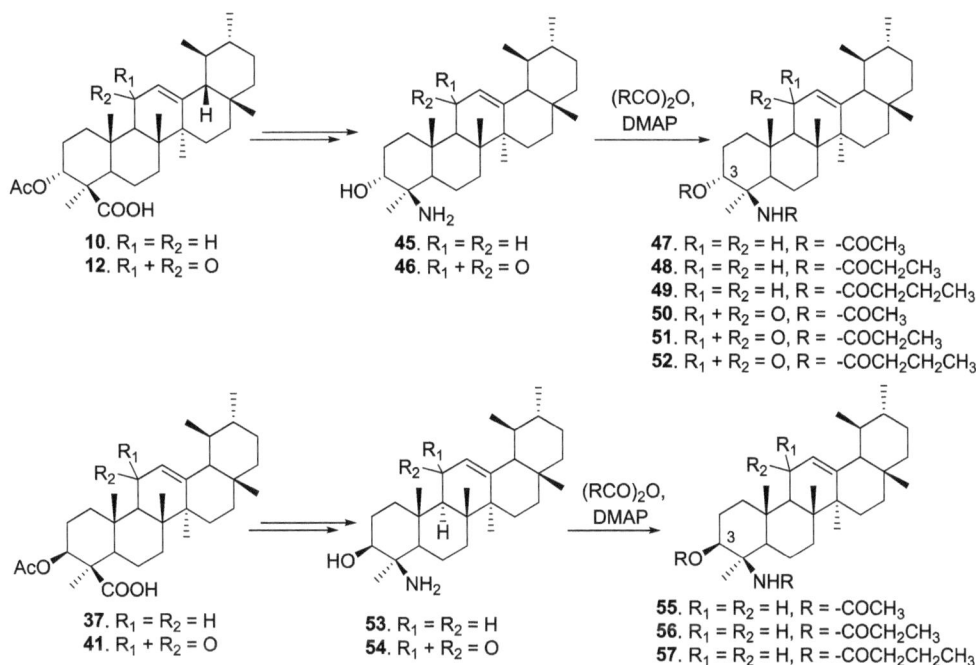

10. $R_1 = R_2 = H$
12. $R_1 + R_2 = O$

45. $R_1 = R_2 = H$
46. $R_1 + R_2 = O$

47. $R_1 = R_2 = H$, R = -$COCH_3$
48. $R_1 = R_2 = H$, R = -$COCH_2CH_3$
49. $R_1 = R_2 = H$, R = -$COCH_2CH_2CH_3$
50. $R_1 + R_2 = O$, R = -$COCH_3$
51. $R_1 + R_2 = O$, R = -$COCH_2CH_3$
52. $R_1 + R_2 = O$, R = -$COCH_2CH_2CH_3$

37. $R_1 = R_2 = H$
41. $R_1 + R_2 = O$

53. $R_1 = R_2 = H$
54. $R_1 + R_2 = O$

55. $R_1 = R_2 = H$, R = -$COCH_3$
56. $R_1 = R_2 = H$, R = -$COCH_2CH_3$
57. $R_1 = R_2 = H$, R = -$COCH_2CH_2CH_3$

Scheme 2. 4-amino derivatives of BAs.

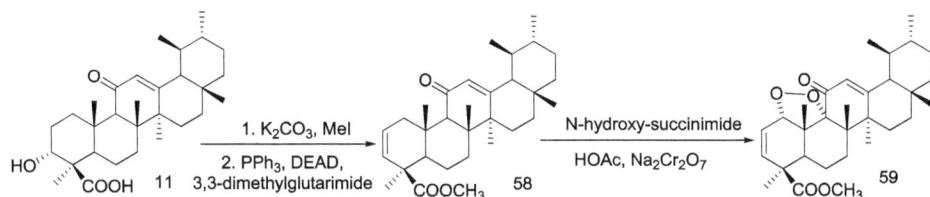

Scheme 3. Endoperoxide of 11-keto-β-boswellic acid.

In another attempt by our group, 2-cyano-1-en-3-one derivatives of β-BAs and 11-keto-β-BAs have been synthesized (Scheme **4**) [129]. Initially, **9** and **11** were successfully transformed into butyl 2-cyano-β-boswellic butyl ester (**60**) and 2-cyano-11-keto-β-boswellic butyl ester (**61**). The treatment of **60** and **61** with DDQ afforded the required butyl 2-cyano-3-oxours-1,12-dien-24-oate **62** and butyl 2-cyano- 3,11-dioxours-1,12-dien-24-oate **63** respectively [130]. Furthermore, the intermediates **60** and **61** were successfully transformed to 4-amino-3-hydroxy-urs-12-en-2-carboxamide **66** and 4-amino-3-hydroxy-11-oxours-12-en-2- -carbox-amide **67** in two steps. Compounds were screened for cytotoxicity against two human cancer cell lines *viz.*, HL-60 and HeLa cells. The cyanoenone **62** and **63** analogues displayed higher cytotoxicity than their parent molecules with IC_{50} values 0.42 and 0.67 μM respectively against HL-60 cell line. The cyano analogues **60** and **61** lacking 1-ene functionality were found to be less active than **62** and **63**. The corresponding amide derivative **66** and **67** were also found to have low cytotoxicity as evidenced by their high IC_{50} values 37 and >50 μM respectively against HL-60 cell line.

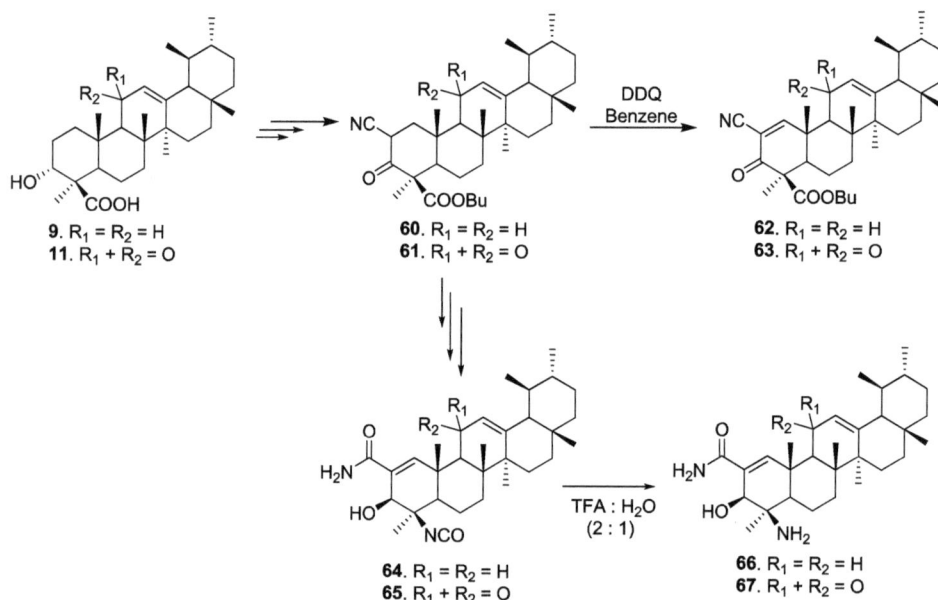

Scheme 4. 2-cyano-1-en-3-one derivatives of BAs.

Recently, Shen and co-workers [131] synthesized different heterocyclic analogues of 11-keto-β-boswellic acid (indole **69**, isoxazoles **71**, and pyrazole derivatives **73-78**) (Scheme **5**). The synthesized analogues were screened for anti-proliferation activities against HUVECs cancer cells, wherein pyrazole derivative (**73**, IC_{50} = 2.13 μM) displayed greater potency than indole (**69**, IC_{50} = 8.22 μM) and isoxazole (**71**, IC_{50} = 9.92 μM) derivatives. They further investigated the anti-proliferation activities of different substituents at nitrogen atom in pyrazole substituted AKBA derivative **73** and observed that the protections with different substituents (acetyl **74**, benzoyl **75**, ethylcarbamoyl **76**, phenylcarbamoyl **77** and isobutoxycarbonyl **78**) resulted in loss of activities against HUVECs cancer cells.

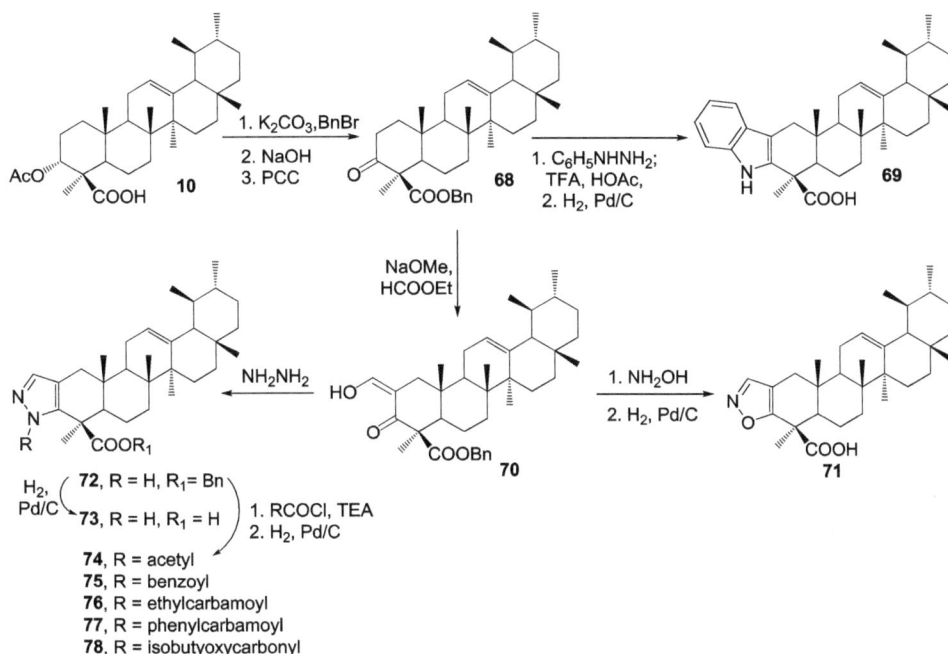

Scheme 5. Indole, isoxazoles and pyrazole derivatives of AKBA.

Csuk and co-workers [132] synthesized ring A modified 11-keto-β-boswellic acid derivatives and evaluated these derivatives against different cancer cell lines. The treatment of 11-keto-β-boswellic acid with $Na_2Cr_2O_7$ afforded the decarboxylated product **79**, which on treatment with excess *m*-CPBA furnished lactone **80**. The reduction of 11-keto-β- boswellic acid with $LiAlH_4$ provided the diol **81**. In addition, the other ring A modified 11-keto-β- boswellic acid derivatives has been synthesised **82-86** (Scheme **6**). These analogues were evaluated for antitumor activity against various cancer cell lines. Most of the synthesized compounds displayed a moderate to good IC_{50} values in the different cancer cells. The biological evaluation revealed that decarboxylated **79** and the ring contracted derivative **82** exhibited a diminished cytotoxicity, suggesting the importance of intact ring A. Furthermore, an amino group at position C-2 **83** led to a significantly improved cytotoxic activity with IC_{50} value 5.6 μM against A253 cancer cells. The analogue having additional -COOH group **84** caused loss of cytotoxicity. The other modifications led to moderate change in the cytotoxicity. The lead compound **83** was found to induce apoptosis in A2780 cancer cells determined using DNA laddering and trypan blue staining experiments.

Ring A modified other structures

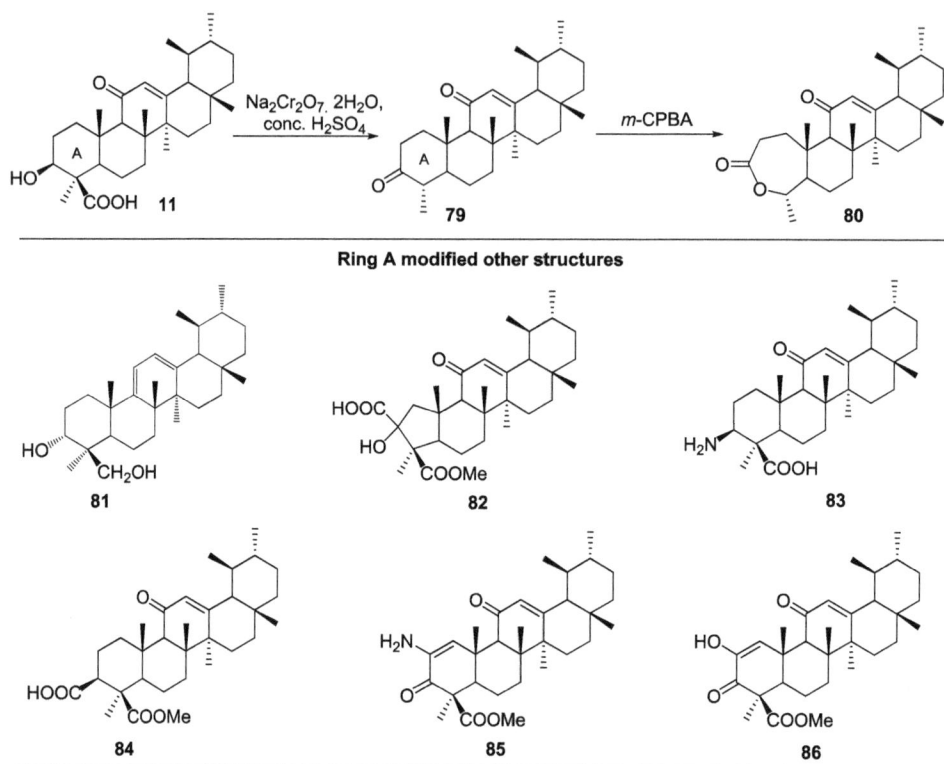

Scheme 6. Ring A modified 11-keto-β-boswellic acid derivatives.

Li and co-workers [133] reported ring A modified derivatives of acetyl-11-ket-
-β-boswellic acid (AKBA). The compound **12** was transformed into diol **87** by
alkali hydrolysis, followed by reduction. The compound **87** was subjected to
selective silyl protection of primary alcohol, followed by oxidation of secondary
alcohol and silyl deprotection to provide keto derivative. Then, Baeyer-Villiger
oxidative rearrangement was accomplished using *m*-CPBA to get the product **88**.
The protection of alcohol with different acylation reagents afforded the
derivatives **89-92**. Furthermore, the compound **88** was oxidised with PDC to
afford aldehyde **93,** which was transformed into compound **94** by a two-step
sequence. The treatment of aldehyde with TBHP and CuCl afforded the free acid
95, which was subjected to esterification with different alcohols to yield **96-99**.
The amide derivatives **100-104** were also synthesized by coupling of acids with
different amines (Scheme **7**). The synthesized derivatives were screened for
anticancer activity against two cancer cell lines (PC-3 and A549). These
derivatives displayed moderate to good cytotoxicity. The **98**, **1022** and methyl
glycine **104** derivatives displayed better cytotoxicity than AKBA in PC-3 cancer
cell lines with IC$_{50}$ values 9, 9 and 8 µM respectively. The other derivatives were
found to be have lower cytotoxicity than parent molecule **12**.

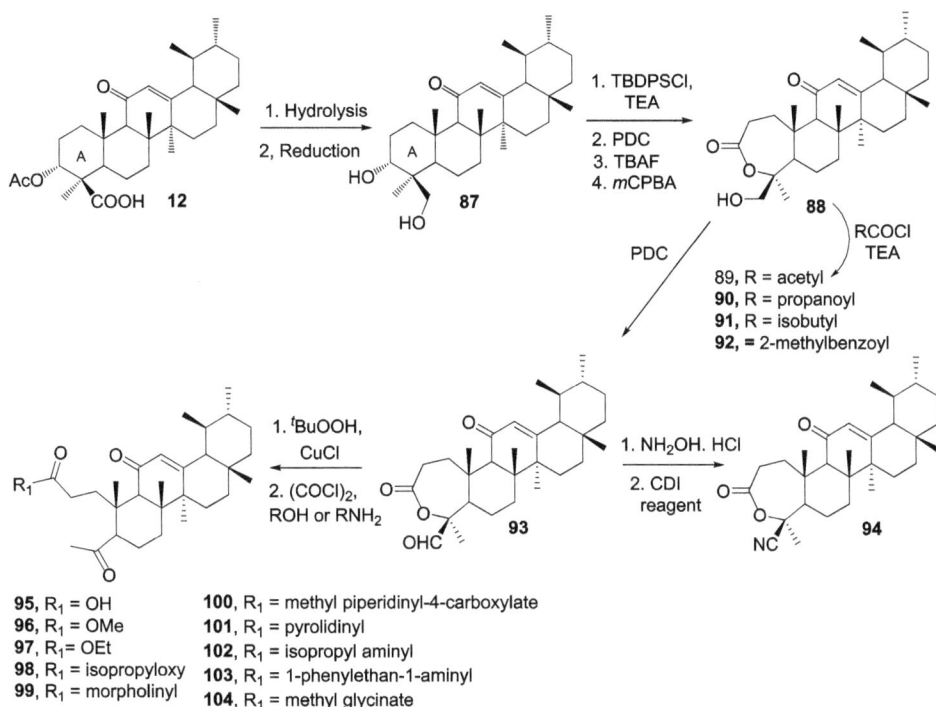

Scheme 7. Ring A modified 11-keto-β-boswellic acid derivatives by Li and co-workers.

The diol derivatives of BAs has also been isolated as well as synthesized. The diol derivative **105** was synthesized by reduction of boswellic acid with LiAlH$_4$, whereas LiAlH$_4$ reduction of 11-keto-boswellic acid afforded the diol derivative **81** (Scheme **8**). The diene-diol **81** showed low IC$_{50}$ values against several human tumour cell lines (*e.g.* DLD-1 = 19.5 μM, A-253 = 18.1 μM and HT-29 = 23.9 μM). The diol derivative **105** has also shown anti-cancer activity *in vivo* models and induced apoptosis in HL-60 cells.

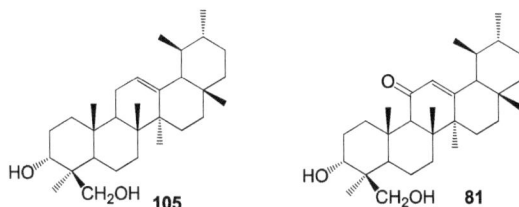

Scheme 8. The diol derivative of BAs.

Shenvi and co-workers synthesized boswellic acid-NSAID (nonsteroidal anti-inflammatory drugs) hybrid molecules (Scheme 9). The methyl esters **106** and **107** were coupled with the various anti-inflammatory drugs (ibuprofen, naproxen, diclophenac and indomethacin) using steglich esterification conditions. The synthesized hybrids **108-115** were evaluated for the anti-inflammatory and anti-arthritic activity. The hybrids **108**, **109**, **112** and **115** exhibited pronounced antiarthritic activity. It was observed that hybrid molecules **108** and **112** have more effectively inhibited *in vivo* COX-2 than ibuprofen drug. Furthermore, hybrids **108** and **112** were also evaluated for *in vitro* lipoxygenase and cyclooxygenase-2 inhibition activity and studies revealed that both hybrids inhibited COX-2 better than LOX enzyme [136].

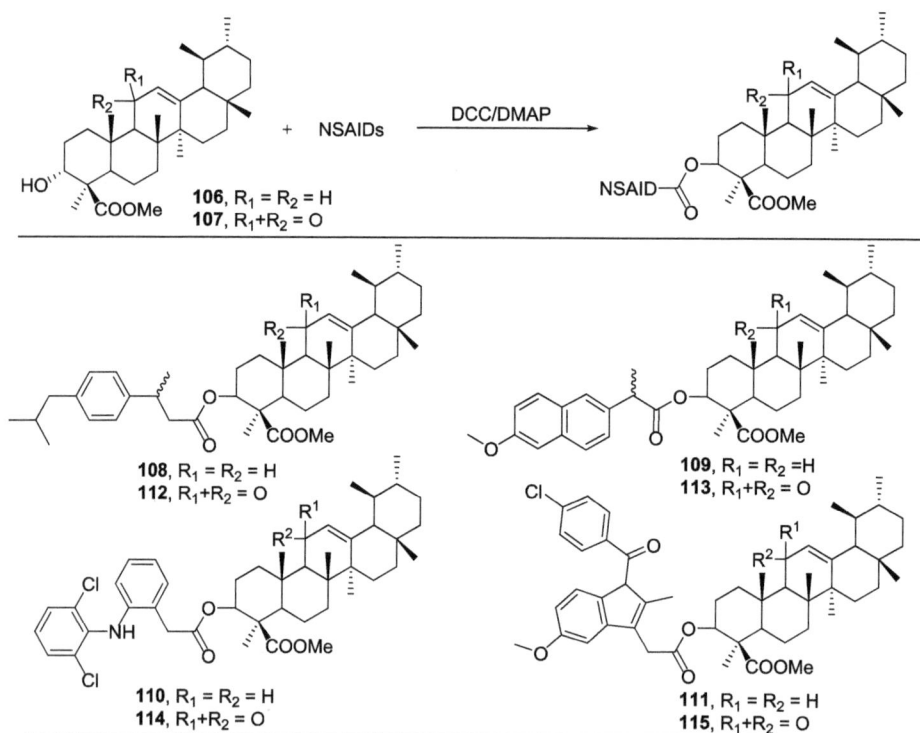

Scheme 9. Synthesis of hybrids of β-boswellic acid and 11-keto-β-boswellic acids.

Chaturvedi and co-workers [137] also synthesized a series of hybrid molecules of β-boswellic acid and 11-keto-β-boswellic acid with anti-inflammatory drugs (aspirin, naproxen, ibuprofen and cinnamic acid). The hybrids were synthesized by the coupling of acid chloride of drugs with **9** and **11** using DMAP as catalyst (Scheme 10). The synthesized hybrids were evaluated for the anti-inflammatory and anti-arthritic activities. The naproxen hybrids with β-boswellic acid **118** and

11-keto-β-boswellic acid **122** exhibited better anti-inflammatory and anti-arthritic activities comparable to that of naproxen.

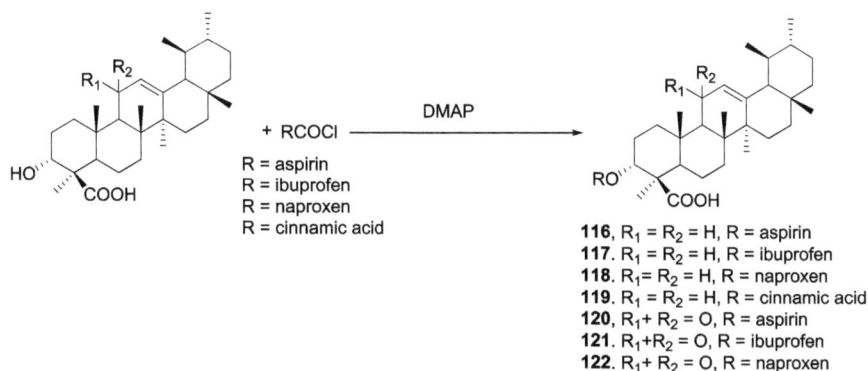

Scheme 10. Hybrids of β-boswellic acid and 11-keto-β-boswellic acid.

Recently, Csuk and co-workers [138] synthesized amide derivatives of 11-keto-β-boswellic acid. The acetyl-11-keto-β-boswellic acid was transformed into acid chloride with oxalyl chloride and consequently reaction with different amines afforded the amide derivatives **123-127**. In addition to amide derivatives, monodesmosidic saponins from 11-keto-β-boswellic acid have been also synthesised **128-129** (Scheme **11**).

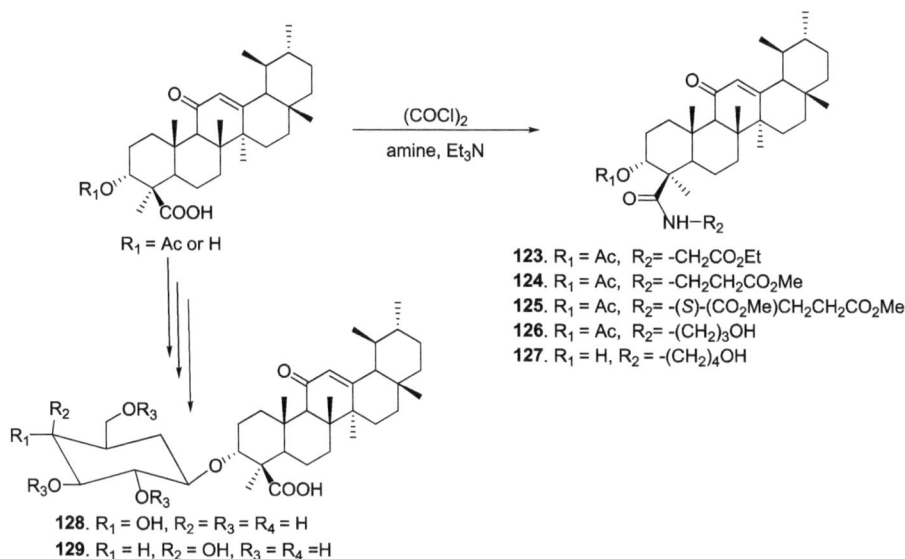

Scheme 11. 11-Keto-β-boswellic acid derived amides and monodesmosidic saponins.

All synthesized compounds were evaluated against a broad panel of human cancer cell lines using photometric sulforhodamine B assay. The bio-evaluation results of amide derivatives revealed that the removal of acetate group at position C-3 decreased the cytotoxicity, whereas amides at C-24 led to enhancement in cytotoxicity. The saponins **128** and **129** were found to be as cytotoxic as AKBA. One of the amide derivative **126** (IC$_{50}$ values 4.5 µM against MCF-7 cancer cells) induced apoptosis in A2780 cells, as validated by DNA laddering experiments and a trypan blue dye exclusion test.

Histone deacetylase (HDAC) inhibitors have recently emerged as promising therapeutic targets for cancer therapy. Generally HDAC inhibitors consists of three important regions: a surface recognition cap group, a hydrophobic linker and zinc-binding functional group. BAs have also been studied for their HDAC inhibitory potential *via.* using boswellic acid core as 'cap' and hydroxamic acid as zinc binding functional group [139]. Synthesis firstly involved the formation of amino BA analogue (**130, 131**), followed by the coupling with hydrophobic linkers *i.e.* monomethyl esters of different carbon chain lengths (**132-139**). In the next step, two types of zinc binding functionalities have been introduced in the molecule *i.e.* carboxylic acid by hydrolysis of ester with 2 N NaOH solution (**140-147**) and hydroxamic acid by treatment with excessive hydroxyl-amine hydrochloride and KOH in methanol (**148-155**) (Scheme **12**). All the compounds possess significant cytotoxicity against a panel of human cancer cell lines. Compound (**144**) has been identified as a lead structure (IC$_{50}$ = 6 µm for HDACs), found to induce G1 cell cycle arrest and caused significant loss in mitochondrial membrane.

In an another attempt by our group [140], biological potential of BAs have been explored by introducing α,β-unsaturated ketone moiety to the core structure. In the first step esterification with diazomethane was done followed by the oxidation of 3-hydroxy group of boswellic acid. In the next step, aldol reaction with aromatic and hetero-aromatic aldehydes have been employed, which subsequently undergo hydrolysis to afford α,β-unsaturated keto boswellic acids (**157-176**) (Scheme **13**). Compounds were evaluated for their anticancer activity against the panel of human cancer lines. One of the lead analog (**173**) displayed significant anticancer activity (1.26 µM against T47D cancer cells) and found to induce apoptosis, confirmed by a series of relevant experiments.

A general pictorial presentation of structure activity relationship of boswellic acids is depicted in Fig. (**8**).

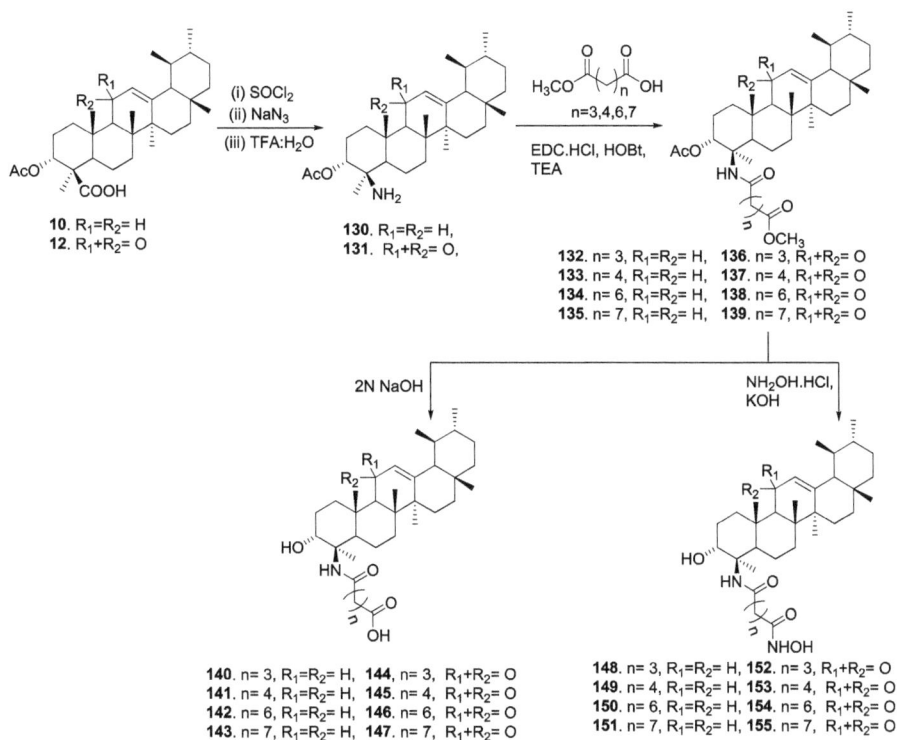

Scheme 12. β-boswellic acid derived HDAC inhibitors.

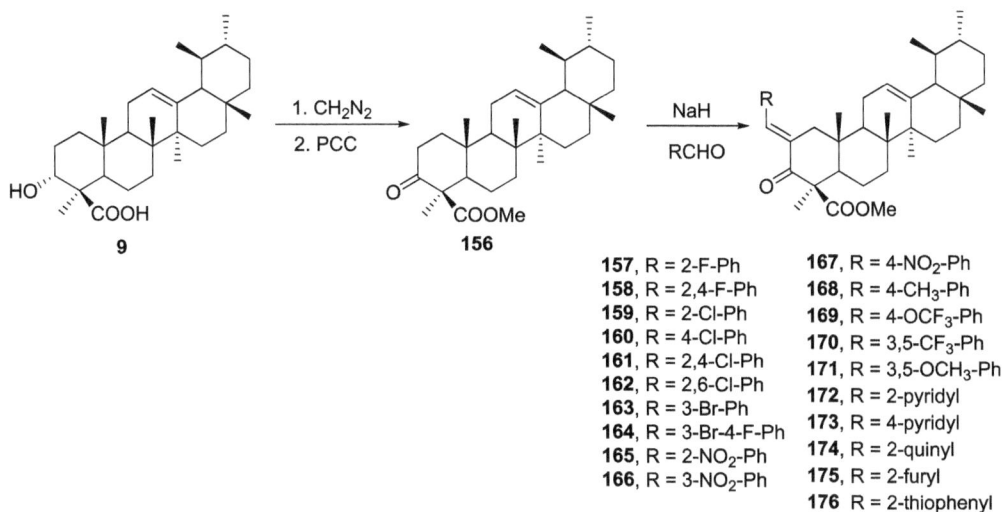

157, R = 2-F-Ph
158, R = 2,4-F-Ph
159, R = 2-Cl-Ph
160, R = 4-Cl-Ph
161, R = 2,4-Cl-Ph
162, R = 2,6-Cl-Ph
163, R = 3-Br-Ph
164, R = 3-Br-4-F-Ph
165, R = 2-NO2-Ph
166, R = 3-NO2-Ph

167, R = 4-NO2-Ph
168, R = 4-CH3-Ph
169, R = 4-OCF3-Ph
170, R = 3,5-CF3-Ph
171, R = 3,5-OCH3-Ph
172, R = 2-pyridyl
173, R = 4-pyridyl
174, R = 2-quinyl
175, R = 2-furyl
176 R = 2-thiophenyl

Scheme 13. β-boswellic acid derivatives based on aldol condensation approach.

Substitution at C-2 position has variable effect on activity.
Double bond at C-1 position enhances the cytotoxicty

Six membered ring is important for activity, as both increase and decrease in ring size reduces the cytotoxicity

α-configuration at C-3 position is important.
Acetyl protection enhances the activity.
3-amino analogs displayed better cyctoxicity than parent molecule.

Reduction of acid group resulted in the loss of cytotoxicity.
Amine and Amide groups at C-4 enhances the acitivity.
Acyl amines at C-4 reduces the activity.
Cyanide group at C-4 reduces the activity,

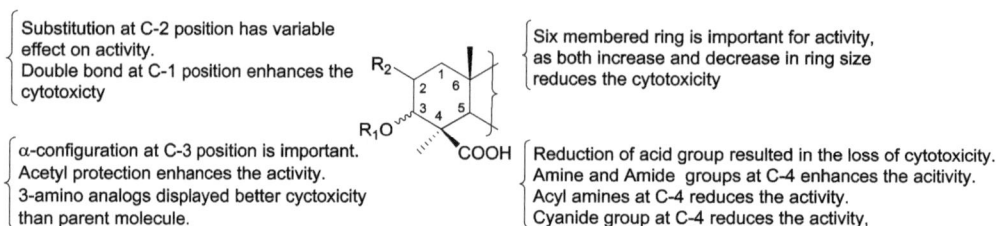

Fig. (8). Structure activity relationship of boswellic acids.

CONCLUSION

Recent SAR studies of BAs have unfolded many facets regarding the biological potential of this class of compounds. The huge biological diversity of these compounds always provides the opportunities to explore them in different possible biological targets. In addition, natural origin of these molecules further opens up the opportunity to bring new isolates with new biological investigations and lot of positive hope in sustainable therapeutics.

CONFLICT OF INTEREST

The authors confirm that they have no conflict of interest to declare for this publication.

ACKNOWLEDGEMENTS

We thank DST (07/2013) and CSIR (BSC-0108) for financial assistance. BAS thanks CSIR, India for Young Scientist Award.

REFERENCES

[1]   Shah BA, Qazi GN, Taneja SC. Boswellic acids: a group of medicinally important compounds. Nat Prod Rep 2009; 26(1): 72-89.
[http://dx.doi.org/10.1039/B809437N] [PMID: 19374123]

[2]   Ammon HP. Boswellic acids in chronic inflammatory diseases. Planta Med 2006; 72(12): 1100-16.
[http://dx.doi.org/10.1055/s-2006-947227] [PMID: 17024588]

[3]   Poeckel D, Werz O. Boswellic acids: biological actions and molecular targets. Curr Med Chem 2006; 13(28): 3359-69.
[http://dx.doi.org/10.2174/092986706779010333] [PMID: 17168710]

[4]   Petronelli A, Pannitteri G, Testa U. Triterpenoids as new promising anticancer drugs. Anticancer Drugs 2009; 20(10): 880-92.
[http://dx.doi.org/10.1097/CAD.0b013e328330fd90] [PMID: 19745720]

[5]   Willmann M, Wacheck V, Buckley J, *et al.* Characterization of NVX-207, a novel betulinic acid-derived anti-cancer compound. Eur J Clin Invest 2009; 39(5): 384-94.
[http://dx.doi.org/10.1111/j.1365-2362.2009.02105.x] [PMID: 19309323]

[6] Fernández Fernández B, Elewa U, Sánchez-Niño MD, *et al.* 2012 update on diabetic kidney disease: the expanding spectrum, novel pathogenic insights and recent clinical trials. Minerva Med 2012; 103(4): 219-34.
[PMID: 22805616]

[7] Khanna D, Sethi G, Ahn KS, *et al.* Natural products as a gold mine for arthritis treatment. Curr Opin Pharmacol 2007; 7(3): 344-51.
[http://dx.doi.org/10.1016/j.coph.2007.03.002] [PMID: 17475558]

[8] Ammon HP. Modulation of the immune system by Boswellia serrata extracts and boswellic acids. Phytomedicine 2010; 17(11): 862-7.
[http://dx.doi.org/10.1016/j.phymed.2010.03.003] [PMID: 20696559]

[9] Yadav VR, Prasad S, Sung B, Kannappan R, Aggarwal BB. Targeting inflammatory pathways by triterpenoids for prevention and treatment of cancer. Toxins (Basel) 2010; 2(10): 2428-66.
[http://dx.doi.org/10.3390/toxins2102428] [PMID: 22069560]

[10] Gilardi D, Fiorino G, Genua M, Allocca M, Danese S. Complementary and alternative medicine in inflammatory bowel diseases: what is the future in the field of herbal medicine? Expert Rev Gastroenterol Hepatol 2014; 8(7): 835-46.
[http://dx.doi.org/10.1586/17474124.2014.917954] [PMID: 24813226]

[11] Du Z, Liu Z, Ning Z, *et al.* Prospects of boswellic acids as potential pharmaceutics. Planta Med 2015; 81(4): 259-71.
[http://dx.doi.org/10.1055/s-0034-1396313] [PMID: 25714728]

[12] Maupetit P. New Constituents in Olibanum Resinoid and Essential Oil. Perfum Flavor 1984; 9: 19-37.

[13] Leung AY, Foster S. Encyclopedia of common natural ingredients used in food, drugs and cosmetics. 2nd ed., New York: John Wiley and Sons 1996.

[14] Wallis TE. Textbook of Pharmacognosy. 5th ed., London: J and A Churchill Limited 1967.

[15] Evans WC. Trease and Evans Pharmacognosy. 14th ed., London: WB Saunders Company Ltd 1996.

[16] Buvari PG. Wirksamkeit und Unbedenklichkeit der H15 Ayurmedica-Therapie bei chronisch-entzündlichen Erkrankungen 2001.

[17] Tadesse W, Desalegn G, Alia R. Natural gum and resin bearing species in Ethiopia and their potential application. Invest Agrar: Sist Recur For 2007; 16: 211-21.

[18] Tucker AO. Frankincense and myrrh. Econ Bot 1986; 40: 425-33.
[http://dx.doi.org/10.1007/BF02859654]

[19] Coppen JJ. In Non-wood Forest Products 1: Flavours and Fragrances of Plant Origin. Rome: FAO 1995.

[20] Hillson RM. Gold, frankincense and myrrh. J R Soc Med 1988; 81(9): 542-3.
[PMID: 3054109]

[21] Hamidpour R, Hamidpour S, Hamidpour M, Shahlari M. Frankincense (rǔ xiāng; boswellia species): from the selection of traditional applications to the novel phytotherapy for the prevention and treatment of serious diseases. J Tradit Complement Med 2013; 3(4): 221-6.
[http://dx.doi.org/10.4103/2225-4110.119723] [PMID: 24716181]

[22] Cuaz-Pérolin C, Billiet L, Baugé E, *et al.* Antiinflammatory and antiatherogenic effects of the NF-kappaB inhibitor acetyl-11-keto-beta-boswellic acid in LPS-challenged ApoE-/- mice. Arterioscler Thromb Vasc Biol 2008; 28(2): 272-7.
[http://dx.doi.org/10.1161/ATVBAHA.107.155606] [PMID: 18032778]

[23]   Afsharypuor S, Rahmany M. Essential oil constituents of two African Olibanums available in Isfahan Commercial Market. Iran J Pharmacol Sci 2005; 1: 167-70.

[24]   Safayhi H, Rall B, Sailer ER, Ammon HP. Inhibition by boswellic acids of human leukocyte elastase. J Pharmacol Exp Ther 1997; 281(1): 460-3.
[PMID: 9103531]

[25]   Nusier MK, Bataineh HN, Bataineh ZM, Daradka HM. Effect of frankincense (*Boswellia thurifera*) on reproductive system in adult male rat. J Health Sci 2007; 53: 365-70.
[http://dx.doi.org/10.1248/jhs.53.365]

[26]   Zhao W, Entschladen F, Liu H, *et al.* Boswellic acid acetate induces differentiation and apoptosis in highly metastatic melanoma and fibrosarcoma cells. Cancer Detect Prev 2003; 27(1): 67-75.
[http://dx.doi.org/10.1016/S0361-090X(02)00170-8] [PMID: 12600419]

[27]   Mukherjee S, Banerjee AK, Mitra BN. Plant antitumour agents. Ind J Pharma 1970; 32: 48-9.

[28]   Hillson RM. Gold, frankincense and myrrh. J R Soc Med 1988; 81(9): 542-3.
[PMID: 3054109]

[29]   Garrison F H. An introduction to the history of medicine. 4th ed. Philadelphia: WB Saunders Company Ltd 1929; 136.

[30]   Culpeper N. Pharmacopoeia Londinensis or the London dispensatory further adorned by the studies and collections of the Fellows, now living of the said college. 6th ed., London: P. Cole 1659.

[31]   Sharma A, Mann AS, Gajbhiye V, Kharya MD. Phytochemical profile of boswellia serrata: An overview. Phcog Rev 2007; 1: 137-42.

[31]   Gupta I, Gupta V, Parihar A, *et al.* Effects of Boswellia serrata gum resin in patients with bronchial asthma: results of a double-blind, placebo-controlled, 6-week clinical study. Eur J Med Res 1998; 3(11): 511-4.
[PMID: 9810030]

[32]   Michie CA, Cooper E. Frankincense and myrrh as remedies in children. J R Soc Med 1991; 84(10): 602-5.
[PMID: 1744842]

[33]   Qurishi Y, Hamid A, Zargar MA, Singh SK, Saxena AK. Potential role of natural molecules in health and disease: Importance of boswellic acid. J Med Plants Res 2010; 4: 2778-85.

[34]   Paranjpe PS. Boswellia serrata: Indian Medicinal Plants-Forgotten Healers: A Guide to Ayurvedic Herbal Medicine. Delhi: Chaukhamba Sanskrit Pratishthan Publishers 2001.

[35]   Ancient Egypt medicine Available at : http://www.crystalinks.com/egyptmedicine.html

[36]   Abercrombie TJ. Arabia's frankincense trail. Natl Geogr Mag 1985; 168: 474-513.

[37]   Ammon HP. [Boswellic acids (components of frankincense) as the active principle in treatment of chronic inflammatory diseases]. Wien Med Wochenschr 2002; 152(15-16): 373-8.
[http://dx.doi.org/10.1046/j.1563-258X.2002.02056.x] [PMID: 12244881]

[38]   Bhargava GG, Negi JJ, Ghua HR. Studies on the chemical composition of Salai gum. Indian Forestry 1978; 104: 174-81.

[39]   Kumar A, Saxena VK. TLC and GLC studies on the essential oil from B. Serrata leaves. Indian Drugs 1979; 16: 80.

[40]   Gupta VN, Yadav DS, Jain MP, Atal CK. Chemistry and pharmacology of gum resin of Boswellia serrata (salai guggal). Indian Drugs 1987; 24: 1-6.

[41]   Pearson RS, Singh P. Indian Forest Records. Published by Govt. of India. 1918; 6: p. 321.

[42]   Dennis TJ, Billore KV, Mishra KP. Pharmacognostic study of gum- oleoresin of Boswelliaserrata. Bull Med Ethno Bot Res 1980; 1: 353-60.

[43]   Pfeifer M. Der Weihrauch Geschichte, Bedeutung, Verwendung. Regensburg: Verlag Friedrich Pustet 1997.

[44]   Asmuth T. Der Jemen gilt als neue Basis des Terrornetzwerks al-Quaida. Aberdie Menschen haben noch ein ganz anderes Problem. Ihr Land trocknet aus. Der Grund: dieDroge Kat. Fluter - Magazin der Bundeszentrale für politische Bildung Nr. 2010; 37: 37-9.

[45]   Gresh A, Radvanyi J, Rekacewicz P, Samary C, Vidal D. Unstaaten am Horn von Afrika. Le Monde diplomatique (Atlas der Globalisierung): 2009; 154-5.

[46]   Ernst E. Frankincense: systematic review. BMJ 2008; 337: a2813.
[http://dx.doi.org/10.1136/bmj.a2813] [PMID: 19091760]

[47]   Siddiqui MZ. Boswellia serrata, a potential antiinflammatory agent: an overview. Indian J Pharm Sci 2011; 73(3): 255-61.
[PMID: 22457547]

[48]   Singh S, Khajuria A, Taneja SC, Khajuria RK, Singh J, Qazi GN. Boswellic acids and glucosamine show synergistic effect in preclinical anti-inflammatory study in rats. Bioorg Med Chem Lett 2007; 17(13): 3706-11.
[http://dx.doi.org/10.1016/j.bmcl.2007.04.034] [PMID: 17481895]

[49]   Singh S, Khajuria A, Taneja SC, Johri RK, Singh J, Qazi GN. Boswellic acids: A leukotriene inhibitor also effective through topical application in inflammatory disorders. Phytomedicine 2008; 15(6-7): 400-7.
[http://dx.doi.org/10.1016/j.phymed.2007.11.019] [PMID: 18222672]

[50]   Shenvi S, Kiran KR, Kumar K, Diwakar L, Reddy GC. Synthesis and biological evaluation of boswellic acid-NSAID hybrid molecules as anti-inflammatory and anti-arthritic agents. Eur J Med Chem 2015; 98: 170-8.
[http://dx.doi.org/10.1016/j.ejmech.2015.05.001] [PMID: 26010018]

[51]   Ammon HP, Mack T, Singh GB, Safayhi H. Inhibition of leukotriene B4 formation in rat peritoneal neutrophils by an ethanolic extract of the gum resin exudate of Boswellia serrata. Planta Med 1991; 57(3): 203-7.
[http://dx.doi.org/10.1055/s-2006-960074] [PMID: 1654575]

[52]   Schweizer S, von Brocke AF, Boden SE, Bayer E, Ammon HP, Safayhi H. Workup-dependent formation of 5-lipoxygenase inhibitory boswellic acid analogues. J Nat Prod 2000; 63(8): 1058-61.
[http://dx.doi.org/10.1021/np000069k] [PMID: 10978197]

[53]   Ammon HP. Salai Guggal - Boswellia serrata: from a herbal medicine to a specific inhibitor of leukotriene biosynthesis. Phytomedicine 1996; 3(1): 67-70.
[http://dx.doi.org/10.1016/S0944-7113(96)80012-2] [PMID: 23194863]

[54]   Kulkarni RR, Patki PS, Jog VP, Gandage SG, Patwardhan B. Treatment of osteoarthritis with a herbomineral formulation: a double-blind, placebo-controlled, cross-over study. J Ethnopharmacol 1991; 33(1-2): 91-5.
[http://dx.doi.org/10.1016/0378-8741(91)90167-C] [PMID: 1943180]

[55]   Singh S, Khajuria A, Taneja SC, Khajuria RK, Singh J, Qazi GN. Boswellic acids and glucosamine show synergistic effect in preclinical anti-inflammatory study in rats. Bioorg Med Chem Lett 2007; 17(13): 3706-11.
[http://dx.doi.org/10.1016/j.bmcl.2007.04.034] [PMID: 17481895]

[56]   Han R. Recent progress in the study of anticancer drugs originating from plants and traditional medicines in China. Chin Med Sci J 1994; 9(1): 61-9.
[PMID: 7916218]

[57]   Han R. Highlight on the studies of anticancer drugs derived from plants in China. Stem Cells 1994; 12(1): 53-63.
[http://dx.doi.org/10.1002/stem.5530120110] [PMID: 8142920]

[58]   Huang MT, Badmaev V, Ding Y, Liu Y, Xie JG, Ho CT. Anti-tumor and anti-carcinogenic activities of triterpenoid, beta-boswellic acid. Biofactors 2000; 13(1-4): 225-30.
[http://dx.doi.org/10.1002/biof.5520130135] [PMID: 11237186]

[59]   Winking M, Sarikaya S, Rahmanian A, Jödicke A, Böker DK. Boswellic acids inhibit glioma growth: a new treatment option? J Neurooncol 2000; 46(2): 97-103.
[http://dx.doi.org/10.1023/A:1006387010528] [PMID: 10894362]

[60]   Janssen G, Bode U, Breu H, Dohrn B, Engelbrecht V, Göbel U. Boswellic acids in the palliative therapy of children with progressive or relapsed brain tumors. Klin Padiatr 2000; 212(4): 189-95.
[http://dx.doi.org/10.1055/s-2000-9676] [PMID: 10994549]

[61]   Liu JJ, Nilsson A, Oredsson S, Badmaev V, Zhao WZ, Duan RD. Boswellic acids trigger apoptosis *via* a pathway dependent on caspase-8 activation but independent on Fas/Fas ligand interaction in colon cancer HT-29 cells. Carcinogenesis 2002; 23(12): 2087-93.
[http://dx.doi.org/10.1093/carcin/23.12.2087] [PMID: 12507932]

[62]   Jing Y, Nakajo S, Xia L, *et al.* Boswellic acid acetate induces differentiation and apoptosis in leukemia cell lines. Leuk Res 1999; 23(1): 43-50.
[http://dx.doi.org/10.1016/S0145-2126(98)00096-4] [PMID: 9933134]

[63]   Hoernlein RF, Orlikowsky T, Zehrer C, *et al.* Acetyl-11-keto-beta-boswellic acid induces apoptosis in HL-60 and CCRF-CEM cells and inhibits topoisomerase I. J Pharmacol Exp Ther 1999; 288(2): 613-9.
[PMID: 9918566]

[64]   Azadmehr A, Ziaee A, Ghanei L, *et al.* A randomized clinical trial study: anti-oxidant, anti-hyperglycemic and anti-hyperlipidemic effects of Olibanum Gum in Type 2 Diabetic Patients. Iran J Pharm Res 2014; 13(3): 1003-9.
[PMID: 25276202]

[65]   Bansal N, Mehan S, Kalra S, Khanna D. *Boswellia serrata*-frankincense (A Jesus Gifted Herb); An Updated Pharmacological Profile. Pharmacologia 2013; 4: 457-63.
[http://dx.doi.org/10.5567/pharmacologia.2013.457.463]

[66]   Gupta I, Parihar A, Malhotra P, *et al.* Effects of *Boswellia serrata* gum resin in patients with ulcerative colitis. Eur J Med Res 1997; 2(1): 37-43.
[PMID: 9049593]

[67]   Singh S, Khajuria A, Taneja SC, *et al.* The gastric ulcer protective effect of boswellic acids, a leukotriene inhibitor from *Boswellia serrata*, in rats. Phytomedicine 2008; 15(6-7): 408-15.
[http://dx.doi.org/10.1016/j.phymed.2008.02.017] [PMID: 18424019]

[68]   Ismail SM, Aluru S, Sambasivarao KR, Matcha B. Antimicrobial activity of frankincense of *Boswellia serrata*. Int J Curr Microbiol App Sci 2014; 3: 1095-101.

[69]   Raja AF, Ali F, Khan IA, Shawl AS, Arora DS. Acetyl-11-keto-β-boswellic acid (AKBA); targeting oral cavity pathogens. BMC Res Notes 2011; 4(406): 1-8.
[PMID: 21205301]

[70]   Raja AF, Ali F, Khan IA, *et al.* Antistaphylococcal and biofilm inhibitory activities of acetyl-11-keo-b-boswellic acid from Boswellia serrata. BMC Microbiol 2011; 11(54): 1-9.
[PMID: 21194490]

[71]   Gupta I, Gupta V, Parihar A, *et al.* Effects of Boswellia serrata gum resin in patients with bronchial asthma: results of a double-blind, placebo-controlled, 6-week clinical study. Eur J Med Res 1998; 3(11): 511-4.
[PMID: 9810030]

[72]   Pandey RS, Singh BK, Tripathi YB. Extract of gum resins of Boswellia serrata L. inhibits lipopolysaccharide induced nitric oxide production in rat macrophages along with hypolipidemic property. Indian J Exp Biol 2005; 43(6): 509-16.
[PMID: 15991575]

[73]   Borrelli F, Capasso F, Capasso R, *et al.* Effect of Boswellia serrata on intestinal motility in rodents: inhibition of diarrhoea without constipation. Br J Pharmacol 2006; 148(4): 553-60.
[http://dx.doi.org/10.1038/sj.bjp.0706740] [PMID: 16633355]

[74]   Kamath JV, Asad M. Effect of hexane extract of Boswellia serrata oleo-gum resin on chemically induced liver damage. Pak J Pharm Sci 2006; 19(2): 129-33.
[PMID: 16751123]

[75]   Pandey RS, Singh BK, Tripathi YB. Extract of gum resins of Boswellia serrata L. inhibits lipopolysaccharide induced nitric oxide production in rat macrophages along with hypolipidemic property. Indian J Exp Biol 2005; 43(6): 509-16.
[PMID: 15991575]

[76]   al-Awadi F, Fatania H, Shamte U. The effect of a plants mixture extract on liver gluconeogenesis in streptozotocin induced diabetic rats. Diabetes Res 1991; 18(4): 163-8.
[PMID: 1842751]

[77]   Zutsi U, Rao PG, Kaur S. Mechanism of cholestrol lowering effect of salai guggal *ex Boswellia serrata* Roxb. Indian J Pharmacol 1986; 18: 182-3.

[78]   Mallik A, Goupale D, Dhongade H, Nayak S. Evaluation of *Boswellia serrata* oleo-gum resin for wound healing activity. Pharm Lett 2010; 2: 457-63.

[79]   Ibrahim M, Uddin KZ, Narasu ML. Evaluation of diuretic activity of Boswellia Serrata bark extracts in albino rats. Int J Drug Formulation Res 2011; 2: 179-99.

[80]   Sharma A, Bhatial S, Kharyaz MD, *et al.* Antiinflammatory and analgesic activity of different fractions of *Boswelliaserrata.* Int J Phytomed 2010; 2: 94-9.

[81]   Chopra A, Lavin P, Patwardhan B, Chitre D. Randomized double blind trial of an ayurvedic plant derived formulation for treatment of rheumatoid arthritis. J Rheumatol 2000; 27(6): 1365-72.
[PMID: 10852255]

[82]   Ding Y, Chen M, Wang M, *et al.* Neuroprotection by acetyl-11-keto-β-Boswellic acid, in ischemic brain injury involves the Nrf2/HO-1 defense pathway. Sci Rep 2014; 7002: 1-9.

[83]   Tschirch A. Halbey. Investigations over the secretions: 28. Over the olibanum. Schweiz. Wchschr Pharm 1898; 36: 466-76.

[84]   Winterstein A, Stein G. Saponin series. X. Monohydroxytriterpene acids. Z Phys Chem 1932; 208: 9-25.
[http://dx.doi.org/10.1515/bchm2.1932.208.1-3.9]

[85]   Simpson JC, Williams NE. I. γ-Boswellic acid. J Chem Soc 1938; 686-8.
[http://dx.doi.org/10.1039/JR9380000686]

[86]   Simpson JC, Williams NE. III. The double bond of β-boswellic acid. J Chem Soc 1938; 1712-9.
[http://dx.doi.org/10.1039/JR9380001712]

[87] Ruzicka L, Wirz W. Triterpenes. L. Conversion of β-boswellic acid into α-amyrin. Helv Chim Acta 1939; 22: 948-51.
[http://dx.doi.org/10.1002/hlca.193902201121]

[88] Ruzicka L, Marxer A. Triterpenes. LIII. Conversion of hederagenin into a rearrangement product of α-boswellic acid. Helv Chim Acta 1940; 23: 144-52.
[http://dx.doi.org/10.1002/hlca.19400230118]

[89] Ruzicka L, Wirz W. Triterpenes. LII. Conversion of α-boswellic acid into β-amyrin. Helv Chim Acta 1940; 23: 132-5.
[http://dx.doi.org/10.1002/hlca.19400230115]

[90] Ruzicka L, Wirz W. Triterpenes. LVIII. Preparation of epi-β-amyrin from α-boswellic acid and from β-amyrone. Helv Chim Acta 1941; 24: 248-52.
[http://dx.doi.org/10.1002/hlca.19410240134]

[91] Ruzicka L, Jeger O, Ingold W. Boswellic acid. Helv Chim Acta 1944; 27: 1859-67.
[http://dx.doi.org/10.1002/hlca.194402701236]

[92] Huzii K, Osumi S. Saponins and sterols. XVII. The structure of ursolic acid. Yakugaku Zasshi 1940; 60: 291-300.

[93] Barton DH, Jones ER. Optical rotatory power and structure in triterpenoid compounds. Application of the method of molecular rotation differences. J Chem Soc 1944; 659-65.
[http://dx.doi.org/10.1039/jr9440000659]

[94] Coelho FP. The determination of the structure of the tetracyclic triterpene alcohol, basseol. II. The dehydrogenation of basseol and of bassenol-triterpenes of the β-amyrenol group. Rev. faculdade cienc. Univ Coimbra 1949; 18: 71-140.

[95] Meyer A, Jeger O, Ruzicka L. Zur Kenntnis der Triterpene. 146. Mitteilung. Zur Konstitution der Sojasapogenole C und A. Helv Chim Acta 1950; 33: 672-87.
[http://dx.doi.org/10.1002/hlca.19500330333]

[96] Beton JL, Halsall TG, Jones ER. The chemistry of triterpenes and related compounds. Part XXVIII. β-Boswellic acid. J Chem Soc 1956; 2904-9.
[http://dx.doi.org/10.1039/JR9560002904]

[97] Shamma M, Glick RE, Mumma RO. The nuclear magnetic resonance spectra of pentacyclic triterpenes. J Org Chem 1962; 27: 4512-7.
[http://dx.doi.org/10.1021/jo01059a095]

[98] Pardhy RS, Bhattacharyya SC. Boswellic acid, acetyl-β-boswellic acid, acetyl-11-keto-β-boswellic acid and 11-keto-β-boswellic acid, four pentacyclic triterpene acids from the resin of *Boswellia serrata* Roxb. Indian J Chem 1978; 16B: 176-8.

[99] Lee Y W, Fang Q, Wang Z, Li D, Cook C E. Pentacyclic triterpenoid compounds as topoisomerase inhibitors or cell differentiation inducers. US Patent 5,064,823, 1991.

[100] Hoernlein RF, Orlikowsky T, Zehrer C, et al. Acetyl-11-keto-beta-boswellic acid induces apoptosis in HL-60 and CCRF-CEM cells and inhibits topoisomerase I. J Pharmacol Exp Ther 1999; 288(2): 613-9.
[PMID: 9918566]

[101] Shao Y, Ho CT, Chin CK, Badmaev V, Ma W, Huang MT. Inhibitory activity of boswellic acids from Boswellia serrata against human leukemia HL-60 cells in culture. Planta Med 1998; 64(4): 328-31.
[http://dx.doi.org/10.1055/s-2006-957444] [PMID: 9619114]

[102] Huang MT, Badmaev V, Ding Y, Liu Y, Xie JG, Ho CT. Anti-tumor and anti-carcinogenic activities of triterpenoid, beta-boswellic acid. Biofactors 2000; 13(1-4): 225-30.
[http://dx.doi.org/10.1002/biof.5520130135] [PMID: 11237186]

[103]   Syrovets T, Büchele B, Gedig E, Slupsky JR, Simmet T. Acetyl-boswellic acids are novel catalytic
        inhibitors of human topoisomerases I and IIalpha. Mol Pharmacol 2000; 58(1): 71-81.
        [PMID: 10860928]

[104]   Kumar A, Shah BA, Singh S, *et al.* Acyl derivatives of boswellic acids as inhibitors of NF-κB and
        STATs. Bioorg Med Chem Lett 2012; 22(1): 431-5.
        [http://dx.doi.org/10.1016/j.bmcl.2011.10.112] [PMID: 22123322]

[105]   Chashoo G, Singh SK, Mondhe DM, *et al.* Potentiation of the antitumor effect of 11-keto-β-boswellic
        acid by its 3-α-hexanoyloxy derivative. Eur J Pharmacol 2011; 668(3): 390-400.
        [http://dx.doi.org/10.1016/j.ejphar.2011.07.024] [PMID: 21821018]

[106]   Syrovets T, Gschwend JE, Büchele B, *et al.* Inhibition of IkappaB kinase activity by acetyl-boswellic
        acids promotes apoptosis in androgen-independent PC-3 prostate cancer cells *in vitro* and *in vivo*. J
        Biol Chem 2005; 280(7): 6170-80.
        [http://dx.doi.org/10.1074/jbc.M409477200] [PMID: 15576374]

[107]   Yuan HQ, Kong F, Wang XL, Young CY, Hu XY, Lou HX. Inhibitory effect of acetyl-11-keto-b-
        ta-boswellic acid on androgen receptor by interference of Sp1 binding activity in prostate cancer cells.
        Biochem Pharmacol 2008; 75(11): 2112-21.
        [http://dx.doi.org/10.1016/j.bcp.2008.03.005] [PMID: 18430409]

[108]   Pang X, Yi Z, Zhang X, *et al.* Acetyl-11-keto-beta-boswellic acid inhibits prostate tumor growth by
        suppressing vascular endothelial growth factor receptor 2-mediated angiogenesis. Cancer Res 2009;
        69(14): 5893-900.
        [http://dx.doi.org/10.1158/0008-5472.CAN-09-0755] [PMID: 19567671]

[109]   Takada Y, Ichikawa H, Badmaev V, Aggarwal BB. Acetyl-11-keto-beta-boswellic acid potentiates
        apoptosis, inhibits invasion, and abolishes osteoclastogenesis by suppressing NF-kappa B and NF-
        kappa B-regulated gene expression. J Immunol 2006; 176(5): 3127-40.
        [http://dx.doi.org/10.4049/jimmunol.176.5.3127] [PMID: 16493072]

[110]   Kumar A, Shah BA, Singh S, *et al.* Acyl derivatives of boswellic acids as inhibitors of NF-κB and
        STATs. Bioorg Med Chem Lett 2012; 22(1): 431-5.
        [http://dx.doi.org/10.1016/j.bmcl.2011.10.112] [PMID: 22123322]

[111]   Patil KR, Mohapatra P, Patel HM, *et al.* Pentacyclic triterpenoids inhibit IKKβ mediated activation of
        NF-κB pathway: *in silico* and *in vitro* evidences. PLoS One 2015; 10(5): e0125709.
        [http://dx.doi.org/10.1371/journal.pone.0125709] [PMID: 25938234]

[112]   Liu JJ, Nilsson A, Oredsson S, Badmaev V, Duan RD. Keto- and acetyl-keto-boswellic acids inhibit
        proliferation and induce apoptosis in Hep G2 cells *via* a caspase-8 dependent pathway. Int J Mol Med
        2002; 10(4): 501-5.
        [PMID: 12239601]

[113]   Liu JJ, Nilsson A, Oredsson S, Badmaev V, Zhao WZ, Duan RD. Boswellic acids trigger apoptosis *via*
        a pathway dependent on caspase-8 activation but independent on Fas/Fas ligand interaction in colon
        cancer HT-29 cells. Carcinogenesis 2002; 23(12): 2087-93.
        [http://dx.doi.org/10.1093/carcin/23.12.2087] [PMID: 12507932]

[114]   Xia L, Chen D, Han R, Fang Q, Waxman S, Jing Y. Boswellic acid acetate induces apoptosis through
        caspase-mediated pathways in myeloid leukemia cells. Mol Cancer Ther 2005; 4(3): 381-8.
        [PMID: 15767547]

[115]   Lu M, Xia L, Hua H, Jing Y. Acetyl-keto-beta-boswellic acid induces apoptosis through a death
        receptor 5-mediated pathway in prostate cancer cells. Cancer Res 2008; 68(4): 1180-6.
        [http://dx.doi.org/10.1158/0008-5472.CAN-07-2978] [PMID: 18281494]

[116] Liu JJ, Duan RD. LY294002 enhances boswellic acid-induced apoptosis in colon cancer cells. Anticancer Res 2009; 29(8): 2987-91.
[PMID: 19661305]

[117] Khan S, Kaur R, Shah BA, *et al.* A novel cyano derivative of 11-keto-β-boswellic acid causes apoptotic death by disrupting PI3K/AKT/Hsp-90 cascade, mitochondrial integrity, and other cell survival signaling events in HL-60 cells. Mol Carcinog 2012; 51(9): 679-95.
[http://dx.doi.org/10.1002/mc.20821] [PMID: 21751262]

[118] Morad SA, Schmid M, Büchele B, *et al.* A novel semisynthetic inhibitor of the FRB domain of mammalian target of rapamycin blocks proliferation and triggers apoptosis in chemoresistant prostate cancer cells. Mol Pharmacol 2013; 83(2): 531-41.
[http://dx.doi.org/10.1124/mol.112.081349] [PMID: 23208958]

[119] Pathania AS, Joshi A, Kumar S, *et al.* Reversal of boswellic acid analog BA145 induced caspase dependent apoptosis by PI3K inhibitor LY294002 and MEK inhibitor PD98059. Apoptosis 2013; 18(12): 1561-73.
[http://dx.doi.org/10.1007/s10495-013-0889-4] [PMID: 23948751]

[120] Glaser T, Winter S, Groscurth P, *et al.* Boswellic acids and malignant glioma: induction of apoptosis but no modulation of drug sensitivity. Br J Cancer 1999; 80(5-6): 756-65.
[http://dx.doi.org/10.1038/sj.bjc.6690419] [PMID: 10360653]

[121] Nandan CD, Reshmi P, Uthaman S, *et al.* Therapeutic properties of boswellic acid nanoparticles in prostate tumor-bearing BALB/c mice model. J Nanopharma Drug Delivery 2013; 1: 30-7.
[http://dx.doi.org/10.1166/jnd.2013.1009]

[122] Snima KS, Nair RS, Nair SV, Kamath CR, Lakshmanan VK. Combination of anti-diabetic drug metformin and boswellic acid nanoparticles: a novel strategy for pancreatic cancer therapy. J Biomed Nanotechnol 2015; 11(1): 93-104.
[http://dx.doi.org/10.1166/jbn.2015.1877] [PMID: 26301303]

[123] Khan MA, Singh M, Khan MS, Najmi AK, Ahmad S. Caspase mediated synergistic effect of Boswellia serrata extract in combination with doxorubicin against human hepatocellular carcinoma. BioMed Res Int 2014; 2014: 294143.
[http://dx.doi.org/10.1155/2014/294143] [PMID: 25177685]

[124] Toden S, Okugawa Y, Buhrmann C, *et al.* Novel evidence for curcumin and boswellic acid-induced chemoprevention through regulation of miR-34a and miR-27a in colorectal cancer. Cancer Prev Res (Phila) 2015; 8(5): 431-43.
[http://dx.doi.org/10.1158/1940-6207.CAPR-14-0354] [PMID: 25712055]

[125] Takahashi M, Sung B, Shen Y, *et al.* Boswellic acid exerts antitumor effects in colorectal cancer cells by modulating expression of the let-7 and miR-200 microRNA family. Carcinogenesis 2012; 33(12): 2441-9.
[http://dx.doi.org/10.1093/carcin/bgs286] [PMID: 22983985]

[126] Shen Y, Takahashi M, Byun HM, *et al.* Boswellic acid induces epigenetic alterations by modulating DNA methylation in colorectal cancer cells. Cancer Biol Ther 2012; 13(7): 542-52.
[http://dx.doi.org/10.4161/cbt.19604] [PMID: 22415137]

[127] Shah BA, Kumar A, Gupta P, *et al.* Cytotoxic and apoptotic activities of novel amino analogues of boswellic acids. Bioorg Med Chem Lett 2007; 17(23): 6411-6.
[http://dx.doi.org/10.1016/j.bmcl.2007.10.011] [PMID: 17950603]

[128] Csuk R, Niesen-Barthel A, Barthel A, Kluge R, Ströhl D. Synthesis of an antitumor active endoperoxide from 11-keto-beta-boswellic acid. Eur J Med Chem 2010; 45(9): 3840-3.
[http://dx.doi.org/10.1016/j.ejmech.2010.05.036] [PMID: 20538386]

[129] Kaur R, Khan S, Chib R, *et al*. A comparative study of proapoptotic potential of cyano analogues of boswellic acid and 11-keto-boswellic acid. Eur J Med Chem 2011; 46(4): 1356-66.
[http://dx.doi.org/10.1016/j.ejmech.2011.01.061] [PMID: 21334793]

[130] Subba Rao GS, Kondaiah P, Singh SK, Ravanan P, Sporn MB. Chemical modifications of natural triterpenes - glycyrrhetinic and boswellic acids: evaluation of their biological activity. Tetrahedron 2008; 64(51): 11541-8.
[http://dx.doi.org/10.1016/j.tet.2008.10.035] [PMID: 20622928]

[131] Shen S, Xu X, Liu Z, Liu J, Hu L. Synthesis and structure-activity relationships of boswellic acid derivatives as potent VEGFR-2 inhibitors. Bioorg Med Chem 2015; 23(9): 1982-93.
[http://dx.doi.org/10.1016/j.bmc.2015.03.022] [PMID: 25819335]

[132] Csuk R, Niesen-Barthel A, Schäfer R, Barthel A, Al-Harrasi A. Synthesis and antitumor activity of ring A modified 11-keto-β-boswellic acid derivatives. Eur J Med Chem 2015; 92: 700-11.
[http://dx.doi.org/10.1016/j.ejmech.2015.01.039] [PMID: 25618017]

[133] Li T, Lou H, Fan P, Ye Y. Method for preparation of 3-acetyl-11-keto-β-Boswellic acid (AKBA) structure modified compound and application as antitumor agents, Faming Zhuanli Shenqing. CN 104558096 A 20150429, 2015.

[134] Bunn PA Jr, Keith RL. The future of cyclooxygenase-2 inhibitors and other inhibitors of the eicosanoid signal pathway in the prevention and therapy of lung cancer. Clin Lung Cancer 2002; 3(4): 271-7.
[http://dx.doi.org/10.3816/CLC.2002.n.012] [PMID: 14662036]

[135] Bhushan S, Kumar A, Malik F, *et al*. A triterpenediol from Boswellia serrata induces apoptosis through both the intrinsic and extrinsic apoptotic pathways in human leukemia HL-60 cells. Apoptosis 2007; 12(10): 1911-26.
[http://dx.doi.org/10.1007/s10495-007-0105-5] [PMID: 17636381]

[136] Shenvi S, Kiran KR, Kumar K, Diwakar L, Reddy GC. Synthesis and biological evaluation of boswellic acid-NSAID hybrid molecules as anti-inflammatory and anti-arthritic agents. Eur J Med Chem 2015; 98: 170-8.
[http://dx.doi.org/10.1016/j.ejmech.2015.05.001] [PMID: 26010018]

[137] Chaturvedi D, Dwivedi PK, Chaturvedi AK, Mishra N, Siddiqui HH, Mishra V. Semisynthetic hybrids of boswellic acids: a novel class of potential anti-inflammatory and anti-arthritic agents. Med Chem Res 2015; 24: 2799-812.
[http://dx.doi.org/10.1007/s00044-015-1331-y]

[138] Csuk R, Barthel-Niesen A, Barthel A, Schäfer R, Al-Harrasi A. 11-Keto-boswellic acid derived amides and monodesmosidic saponins induce apoptosis in breast and cervical cancers cells. Eur J Med Chem 2015; 100: 98-105.
[http://dx.doi.org/10.1016/j.ejmech.2015.06.003] [PMID: 26073487]

[139] Sharma S, Ahmad M, Bhat JA, *et al*. Design, synthesis and biological evaluation of β-boswellic acid based HDAC inhibitors as inducers of cancer cell death. Bioorg Med Chem Lett 2014; 24(19): 4729-34.
[http://dx.doi.org/10.1016/j.bmcl.2014.08.007] [PMID: 25176189]

[140] Kumar A, Qayum A, Sharma PR, Singh SK, Shah BA. Synthesis of β-boswellic acid derivatives as cytotoxic and apoptotic agents. Bioorg Med Chem Lett 2016; 26(1): 76-81.
[http://dx.doi.org/10.1016/j.bmcl.2015.11.027] [PMID: 26608550]

CHAPTER 4

Natural Compounds: Cancer Preventive Agents

Sandeep K Misra*

University of Mississippi, Oxford, Mississippi, USA 38677

Abstract: Cancer chemoprevention is a rapidly evolving scientific research area. Cancer chemoprevention is the application of natural or synthetic agents to reduce or delay the onset of cancer. Different approaches are currently being used for cancer prevention and one of these important approaches is the use of natural compounds. Several natural compounds are currently under the investigation for their efficacy in preventing cancer. These compounds include curcumin, tea polyphenols, resveratrol, genistein, luteolin, lycopene among others. Curcumin is probably the most studied natural molecule for its ability to prevent cancer. Many of these compounds are being investigated using *in vitro* as well as animal models. Clinical trials are also underway for many natural compounds to test their efficiency in cancer chemoprevention. Recent evidence suggests that these natural molecules have antioxidant, anti-inflammatory, immune-enhancing, cell cycle modifying and cell differentiating, apoptotic and suppression of proliferation and angiogenesis properties. These natural compounds target several molecular pathways. Some of these pathways include p53 family, activator protein 1, NF-κB, growth factors among others.

Keywords: Cancer chemoprevention, Molecular pathways, Natural compounds.

INTRODUCTION

The cancer is the second leading cause of death in the USA [1]. Cancer patients apply all the possible efforts to fight the disease, and manage its symptoms. However, numerous side effects have been found to be associated with currently available treatment regimens. Thus, it's better to prevent the disease than to treat it. Several approaches are suggested that may prevent the onset of cancer. These approaches include changes in the dietary intake, taking nutritional supplements, lifestyle modifications, like exercise and reducing exposure to the sun and decreasing exposure to the polluted environment [2].

Cancer chemoprevention is a way to prevent cancer from originating, or slow

---

* **Corresponding author Sandeep K. Misra**: University of Mississippi, Mississippi, USA 38677; Tel: +1-662-915-2207; Fax: +1-662-915-5118; E-mail: sandeepkmisra@gmail.com

Sahdeo Prasad & Amit Kumar Tyagi (Eds.)

down the progression, or reverse the cancer disease by the administration of one or combination of a few natural and/or synthetic chemical agents.

Research on cancer prevention is an emerging field now and there is not a lot of information available for the ways to prevent the onset of cancer. Theories of cancer prevention generally originate from correlations, observation of populations, and lifestyle as well as diet patterns at the national or cultural level. These factors may affect the rate of onset of cancer. More and more people in the USA and worldwide are leaning towards the complementary and alternative medicine for preventing and managing the disease. An important category in this group is the use of the natural compounds that prevent and treat cancer. Avoidance of chemical, biologic, and physical agents that cause cancer and the eating diets high in cancer preventive factors is the two most important approaches for cancer prevention. It is estimated that just by modification of diet alone, for *e.g.*, increasing high intake of fruits and vegetables, one can reduce the cancer occurrence by 20%.

CANCER PREVENTION THROUGH NATURAL SOURCES

Cancer preventive agents may include foods, spices, compounds extracted from the plants, or specific nutrients found in nature. Cancer prevention research that employs natural compounds are often studied for their role in specific physiological pathways. Most of the published literature on food chemoprevention focuses on changing the nutrition behavior. These include increasing the intake of fruits, vegetables, and foods made from whole grains. These chemopreventive agents may have various properties that make them suitable candidates. Most of these agents have antioxidant and apoptotic, antimetastatic activities and are able to modulate the immune system and hormones. Some of these candidates that have shown their role in cancer prevention are summarized here.

Curcumin

Curcumin is an extract of the rhizome of the turmeric plant (*Curcuma longa*) and it is a major yellow pigment of turmeric. It is the main curcuminoid of turmeric and exhibit potent antioxidant and anti-inflammatory effects. The curcuminoids are natural phenols that impart the yellow color to turmeric. Curcumin may exist in 1,3-diketo and equivalent enol tautomeric forms, (Fig. **1**). The enol form is more energetically stable in the solid phase and in organic solvents, and the 1,3-diketo form is predominantly present in the water.

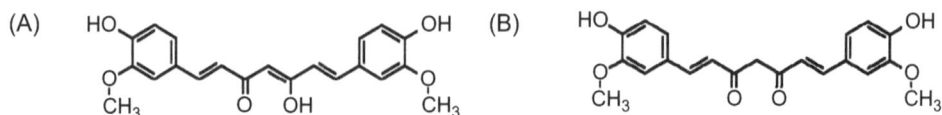

Fig. (1). Structure of curcumin (A) Enol form (B) Keto form.

Turmeric and its extract have been used traditionally as medicine for centuries in treating various symptoms. However, much attention has been paid only recently, for its anticancer activity. Probably, it is the most widely studied natural chemopreventive agent for many cancers, including breast, cervical, colon, gastric, hepatic, leukemia, oral epithelial and ovarian cancer. The role of curcumin in biochemical pathways has been studied in detail. Curcumin targets multiple molecules- that include Ap-1, NF-κB, Akt, STAT3, Bcl-2, Bcl-XL, Poly (ADPribose) polymerase, caspases, Ikappa B kinase, human epidermal growth factor receptor 2, epidermal growth factor receptor, Jun N-terminal Kinase, cyclooxygenase 2, mitogen-activated protein kinases, and 5-lipoxygenase. Curcumin has been reported to induce phase II enzymes, like glutathione S-transferase, as well as inhibit phase I enzymes. Curcumin has been studied in detail in colon cancer models. It has been shown to interrupt the process of carcinogenesis by inhibiting the initiation step or suppressing the promotion and progression steps in the animal models [3, 4]. Curcumin also inhibited the growth of cancer cells *in vitro* and in xenograft models by inducing cell cycle arrest and apoptosis [5 - 7]. Besides these, curcumin also exhibits synergistic che-mopreventive effects with polyphenols found in food, like, genistein [8], green tea [9] and embelin [10]. The activity of several drugs to treat cancer, like, vinorelbine, vinca alkaloid, and fluorouracil is enhanced by the curcumin. More than five phase I clinical trials have been completed for curcumin and establish its tolerability and the safety in patients with colorectal cancer. In these study, up to eight grams curcumin was administered per day [11]. The side effects were very mild and only a few subjects reported mild nausea and diarrhea.

Tea Polyphenols

Tea is the most widely consumed beverages in the world except water. It is the only beverage that is served hot or iced, anytime, anywhere for any occasion. The tea is rich in chemicals that have strong antioxidant activities. Several epidemiological studies from around the world, including *in-vivo* models, indicate that green tea reduces the onset of cancer. Different types are tea are obtained based on processing techniques, including green and black tea. Both these tea types have been investigated for their effectiveness in preventing cancer. Green tea showed higher promise towards various cancer types, as compared to the black tea. The most active compound with anticarcinogenic activity found in the green tea is epigallocatechin-3-gallate (EGCG), which is the most abundant polyphenol in green tea. It is a type of catechin and is an ester of epigallocatechin and gallic acid, (Fig. **2**).

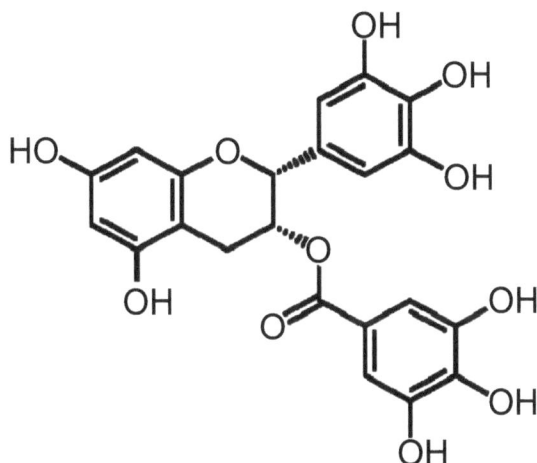

Fig. (2). Structure of epigallocatechin-3-gallate.

EGCG increases the efficacy of some other cancer drugs used for treating cancer *in vitro* and animal models synergistically. It was also effective against the tumor necrosis factor–related apoptosis-inducing ligand resistant prostate cancer cell line apoptosis. In an animal study, polyphenols that were purified from green tea reduced the tumor growth. It also suppressed metastasis of metastasis-specific mouse mammary cancer cells [12]. Green tea polyphenols also caused the reduction in the formation of tumor blood vessels in breast cancer that is estrogen receptor-negative [13]. In another research study, after the rats were challenged with azoxymethane, the green tea polyphenols (GTP) extract reduced the risk of development of colon cancer [14].

Because of some promising *in vitro* and *in vivo* results of green tea in cancer chemoprevention, several clinical trials are being carried out. These trials involve only green tea or they combine other drugs to test their efficacy in preventing cancer. One of these clinical trials is to measure the effect of green tea polyphenols together with erlotinib as an approach for chemoprevention of premalignant lesions of the head and neck (identifier NCT01116336). A case-control study has been performed at the Mayo clinic in patients with chronic lymphocytic leukemia and other low-grade lymphomas. Many of the patients who used products that contain tea polyphenols showed evidence of clinical benefit after from these products [15]. A phase I/II clinical trial has been completed at the Mayo clinic to investigate the role of extracts of decaffeinated green tea in patients with symptomatic, lymphocytic leukemia in early stages. This study was sponsored by the National Cancer Institute (identifier NCT00262743). This study found that the daily oral dose of EGCG reduced the number of lymphocytes count and lymphadenopathy in most of the patients. Several phase I trials have been completed on healthy individuals. These studies helped to define the pharmacokinetic parameters and preliminary safety profiles upon consumption of several preparations. The administration of green tea was relatively very safe among volunteers. One phase I clinical trial suggested that up to 1 g of green tea solids, that is equivalent to approximately 900 mL of green tea, could be safely consumed by patients with solid tumors [16].

Resveratrol

Resveratrol is a stilbenoid, a derivative of stilbene. Resveratrol is a major component of red wine. It is produced by many plants when the plant is damaged or when the pathogens attack these plants. It is a phytoalexin, and food sources of resveratrol include grapes, mulberries, raspberries, and blueberries. Resveratrol is synthesized in plants by the action of the resveratrol synthase enzyme, (Fig. **3**).

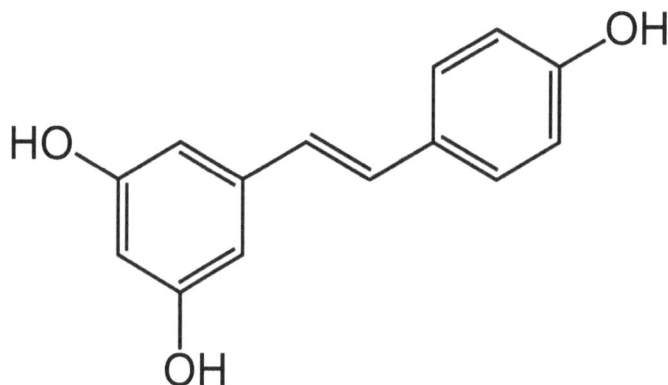

Fig. (3). Structure of the resveratrol.

There are some studies that suggest cardioprotective and chemopreventive activities of resveratrol. In one study, resveratrol prevented the development of skin cancer in mice that were treated with the carcinogen. Resveratrol was also found to be efficient in all the three major stages of cancer in the same study [17]. Decreased skin cancer was observed for the mice that were treated with resveratrol before and after the exposure to the ultraviolet light [18]. Reduction in the number and size of esophageal, intestinal, and colon tumors was observed when resveratrol was used as a prophylactic agent [19]. It has been shown that resveratrol can prevent the development of DMBA-induced mammary carcinogenesis. It can also inhibit the growth of xenografts, induce apoptosis of prostate cancer cell lines and suppresses the progression of prostate cancer in transgenic adenocarcinoma of the mouse prostate mice.

Some preclinical studies found resveratrol to be effective against various cancers [20, 21]. Some completed clinical trials of resveratrol (identifier NCT00256334; NCT00098969) found the chemopreventive role of resveratrol. Another clinical trial in underway to assess the role of resveratrol in chemoprevention in colon cancer (identifier NCT00578396).

Lycopene

Lycopene is a bright red antioxidant carotene that gives the red color to the tomatoes, pink grapefruits, watermelon, and many other fruits and vegetables. The lycopene is insoluble in water. Just like all the other carotenoids, lycopene is a polyunsaturated hydrocarbon. Structurally, lycopene is a tetraterpene and eight isoprene units are assembled. These units have only carbon and hydrogen, (Fig. **4**). Eleven conjugated double bonds present in lycopene are responsible for proving it the red color as well as its antioxidative property.

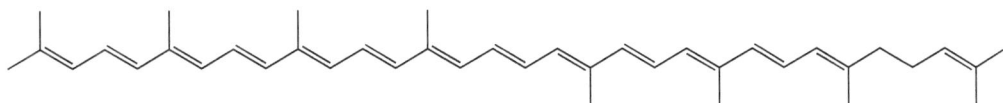

Fig. (4). Structure of Lycopene.

When the rats were administered diets that contain broccoli, tomato, lycopene, and a combination of tomato plus broccoli, a reduction in Dunning R-3327H prostate cancer growth rate was observed [22]. Some epidemiological studies have shown that the risk of specific cancers including the digestive tract, cervix, and prostate cancers can be decreased by high consumption of lycopene-containing vegetables [23, 24]. One study found that tomato products may help benign prostate hyperplasia by preventing cancer progression [25]. It can also help in increasing apoptosis in benign prostate hyperplasia and carcinoma [26]. In another study, a combination of vitamin E, lycopene, and selenium significantly inhibited the development of prostate cancer and increased disease-free survival among the patients with these cancers. In a cell culture model, lycopene strongly inhibited proliferation and induced apoptosis of prostate and breast cancer cell lines. Treatment with lycopene A also causes the reduction in the incidence of aberrant crypt foci. A phase II randomized clinical trial of lycopene as a dietary supplement before radical prostatectomy suggested that lycopene may decrease the growth of the prostate cancer [27]. These reports together suggest that lycopene is helpful in preventing different cancers. In another study, lycopene strongly inhibited the lung cancer cell growth and was reported to be more effective than either α-carotene or β-carotene. Lycopene administration during the postinitiation stage reduced the incidence of lung adenocarcinoma in an animal model. In two large cohort studies, a lower risk of lung cancer was observed to be significantly associated with α-carotene and lycopene intake. Tomato power and lycopene supplementation in the diet prevented leiomyoma of the oviduct in the Japanese quail [28].

Pomegranate

Pomegranate is one of the earliest known cultivated fruits and is used in baking, juice blends, smoothies and cooking, and alcoholic beverages. Pomegranate fruit has been being used for several medicinal properties for a long time, but its chemopreventive property has been investigated in recent times only. Since then, it has drawn much attention as the chemopreventive agent. Several *in vitro* and *in vivo* observations strongly suggest that pomegranate can be a good candidate with its chemoprevention properties against prostate cancer. In one study, the fractions rich in polyphenol from the fermented pomegranate juice, aqueous extract of the pericarp, or supercritical CO_2 extracted seed oil were tested for their anticancer property. These also inhibited the growth of breast cancer cells [29] and decreased the formation of new blood vessels in the chicken chorioallantoic membrane model *in vivo* [30]. It has also been reported that the pomegranate constituents delphinidin, cyanidin, and petunidin inhibited the growth of breast cancer cells [31]. Oil extracted from pomegranate seeds inhibited skin tumor development and promotion in CD1 mice. Treatment with pomegranate fruit extract was shown in one study to be effective in inducing cell cycle arrest and apoptosis of human lung carcinoma cells. Tumor growth in the nude mice was significantly inhibited upon oral administration of extract prepared from pomegranate [32]. It also protected A/J mice from lung carcinogenesis, that was induced by the carcinogen [33]. Extract of pomegranate fruit was reported to inhibit the growth of prostate cancer lines. It induced the apoptosis and inhibited xenografts as well as decrease in serum prostate-specific antigen (PSA) levels [34]. This activity of the pomegranate fruit extract was dose dependent. The proliferation of LNCaP and human umbilical vein endothelial cells were found to be significantly inhibited by pomegranate extract. When the pomegranate extract was administered to severe combined immunodeficiency mice, the decrease in size for xenografted prostate cancer, vascular endothelial growth factor (VEGF) peptide levels, reduced HIF-1α expression and the decrease in tumor vessel density was observed [35].

A phase II clinical trial, to assess the effects of pomegranate juice consumption on PSA progression in men with rising PSA after primary surgery, or radiation showed a significant increase in mean PSA doubling time [36]. The statistically significant prolongation of PSA doubling time, coupled with corresponding laboratory effects on prostate cancer cell proliferation and apoptosis, encourages the further studies of these effects.

Luteolin

Luteolin is a flavone, which is a type of flavonoid. It is found in many green vegetables, such as cabbage, broccoli, celery, green pepper, spinach, and cauli-flower, (Fig. **5**). Luteolin displays a wide spectrum of pharmacological activities. These activities range from anti-inflammatory to anticancer properties [37]. The observed biological effects of luteolin in chemoprevention could be functionally related to other properties of luteolin. For *e.g.*, the anti-inflammatory activity exhibited by luteolin may be linked to its anticancer property.

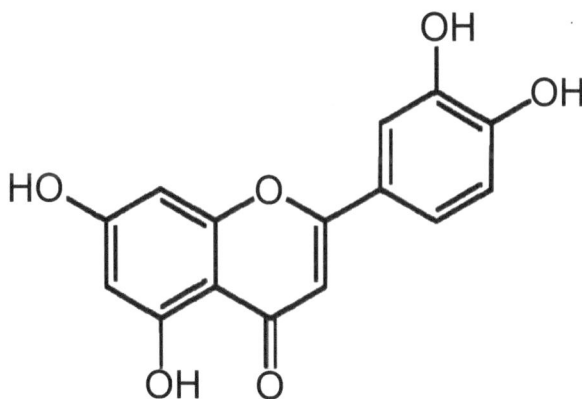

Fig. (5). Structure of the Luteolin.

Luteolin sensitizes cancer cells to therapeutic-induced cytotoxicity by the suppression of cell survival pathways. These include nuclear factor kappa B, phosphatidylinositol 3'-kinase, stimulating apoptosis pathways and X-linked inhibitor of apoptosis protein, and including pathways that induce p53, a tumor suppressor. These observations suggest that luteolin could be used as an anticancer agent for several cancer types. Anticancer activity expressed by Luteolin is mediated by inducing senescence, cell cycle arrest, or apoptosis in oral squamous cancer cells [38], lung carcinoma cells [39], human esophageal adenocarcinoma cells [40], human colon cancer cells, and human hepatoma cells [41]. It also induced apoptosis *in vitro* in the prostate cancer cell lines. Efficacy of cisplatin in gastric cancer cells was increased by luteolin. DMBA-induced mammary tumors were shown to be inhibited by luteolin. The incidence of colon cancer was also found to be significantly decreased by luteolin. The reduction in the number of tumors per rat was also observed when luteolin was administered to the mice at the initiation and the postinitiation stages of the cancer development. Clinical trials with well-defined controls are now required to further investigate the chemopreventive properties of luteolin in healthy volunteers.

Genistein

Genistein is an angiogenesis inhibitor phytoestrogen that is found mainly in soybeans and soy products, (Fig. **6**). The consumption of dietary genistein decreased tumor multiplicity and reduced the incidence of adenocarcinoma in the DMBA model of mammary cancer in rats. The soybean isoflavone mixture that consisted 74% genistein and 21% daidzein inhibited the adenocarcinoma in the prostate and seminal vesicles in rats that were induced by DMBA. Genistein inhibited prostate cancer cell growth in culture by inducing G2/M arrest and apoptosis, inhibited the secretion of prostate specific antigen [42]. It also increased the effect of radiation against prostate cancer in cell culture and in orthotopic and metastatic *in vivo* models [43]. The antitumor activity of cisplatin in BxPC-3 tumor xenografts was also increased by the genistein.

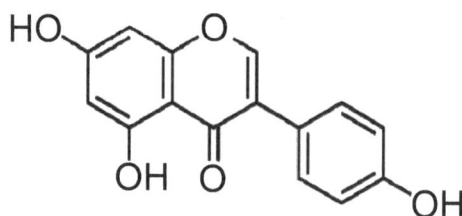

Fig. (6). Structure of Genistein.

Several epidemiological studies provide evidence that high intake of soy consumption reduces the risk of prostate [44], breast [45], and endometrial cancers [46]. Encouraged by these anticancer observations of genistein, many early stage clinical trials have been carried out utilizing either soy products or genistein. One clinical trial study that was conducted in prostate cancer patients with rising serum prostate specific antigen levels suggested that soy isoflavones benefit some prostate cancer patients. Another phase II clinical trial was completed in prostate specific antigen-recurrent prostate cancer patients. After the therapy, a decrease in serum prostate-specific antigen level from 56% to 20% was observed [47]. Few other clinical trials are underway to study the effectiveness of genistein and soy products in cancer prevention (identifier NCT01538316; NCT00099008).

Mushrooms

Many mushrooms have now been identified that can modulate immune system of the human subjects. Three species of mushrooms, Coriolus *versicolor*, Ganoderma *lucidum* and Grifola *frondosa* are particularly important in modulating the immune system. These contain polysaccharides that exhibit immune system modulating effects. These effects have implications in preventing the onset as well as treatment of cancer. Coriolus *versicolor* had been used in traditional Chinese medicine for a long time. It is also often used in Japanese cancer-related treatments. Coriolus *versicolor* is mostly used in extract form, either as protein-bound polysaccharide-K or polysaccharide-P. Recent research has shown immune-stimulatory actions of these polysaccharides on monocytes, macrophages, B-lymphocytes, T-lymphocytes, bone marrow cells, and lymphocyte-activated killer cells. These also promote the production and proliferation of antibodies, interferons, tumor necrotic factor, and different cytokines like, IL-2 and IL-6 [48, 49]. Some studies on Ganoderma *lucidum* have reported its inhibitory effects on AP-1, uPAR, NF-κB, uPA. Many preliminary clinical studies have reported some beneficial effects on the immune system in patients with lung cancer or colon cancer. A mushroom Grifola *frondosa* is used for human consumption and shows anticancer properties and currently under investigation. The most biologically active molecules in this mushroom are polysaccharides 1,3 and 1,6 beta-glucan. These constituents are available in the market in the form of proprietary D- or MD-fraction extracts. The extract from Grifola *frondosa* extracts has been shown to induce apoptosis in prostate cancer cells, enhance bone marrow colony formation *in vitro*. It has also been shown to reduce the effective dose of mitomycin-C in an animal model with tumors and activate natural killer cells in human cancer patients.

Sulforaphane

Sulforaphane is a compound within the isothiocyanate group of organosulfur compounds, (Fig. 7). Sulforaphane was first discovered in broccoli sprouts, which have the highest concentration of sulforaphane in nature. It can be found in many vegetables such as broccoli, cabbage and brussels sprouts kale, collards, kohlrabi, mustard, turnip, radish, and arugula. Young sprouts of broccoli and cauliflower are particularly rich in glucoraphanin. It is produced in the plants by the enzyme myrosinase. This enzyme converts glucoraphanin into sulforaphane in damaged plants, which allows the two compounds to mix and react.

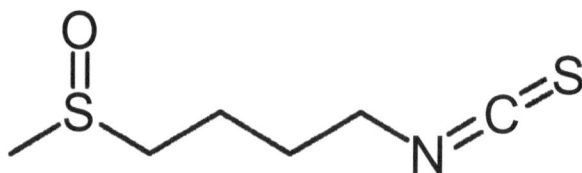

Fig. (7). Structure of Sulforaphane.

It has been shown that to get the actual benefits of sulforaphane, the pharmacological doses in the form of supplement diet would be needed. Dietary sources such as broccoli do not contain sufficient sulforaphane to exert an effect. Studies are needed to establish the safe dose of sulforaphane if high doses are administered. As the more studies are published towards the understanding of the ability of sulforaphane to selectively kill cancer cells, it suggests that it may be a good candidate for the treatment of metastasized cancer. It can be used alongside existing approaches to treating cancer. In some studies, sulforaphane presented toxicity towards various cancer cell lines, like colon, breast, ovarian and prostate cancer [50]. In a mouse model, it was observed that sulforaphane inhibits the metastases of prostate cancer. It has been shown that sulforaphane can change patterns of gene expression as well histone methylation in metastasized prostate cancer cells. Sulforaphane begins a process that can help to re-express tumor suppressors, leading to the selective death of cancer cells and slowing disease progression.

Ginseng

Ginseng belongs to genus Panax and is a slow-growing perennial plant with fleshy roots. The medicinal and chemopreventive effect of ginseng can be attributed to a group of triterpene saponins, called ginsenosides. The concentration of ginsenoside in plant roots depends on the age and type of the plant. Ginseng has promising chemopreventive properties and many ginsenosides have been reported to modulate the host immune system as well as reduce the proliferation of the cancer cells. Many compounds with pharmacological and chemopreventive activity have been purified from Ginseng found in Siberia. Further studies point that the observed benefits of these compounds are because they show protection from and inhibit free radicals. Six compounds purified from the ginseng found in Siberia have a different degree of activity as antioxidants. Four of these compounds were reported to have activity in cancer inhibition. More than one pharmacological effect is observed by some compounds and some compounds show similar effects, although they belong to chemically different class. The most active components of Panax ginseng are ginsenoside saponins. About 30 different types of saponins have been purified from this ginseng. Several of these have been investigated for their potential use in preventing cancer. Examples include ginsenoside Rg3, Rg5, Rh1, and Rh2. Ginseng also possesses anti-inflammatory properties that target many of the important pathways in the inflammation-to-cancer sequence. Preliminary clinical trials have shown that Ginseng improved the quality of life in some cancer patients. An NIH clinical trial is underway to monitor the role of ginseng in controlling fatigue caused by cancer, that may have some implications on cancer care for long-term.

Flaxseed

Flaxseed is rich in fibers, omega-3 fatty acids, and lignans. Flaxseeds exhibit chemopreventive activities in breast cancer as they inhibit the production of the hormone estrogen and these two properties may be linked. Flaxseeds have been reported to reduce the growth of human mammary tumor cells, reduce mammary tumor initiation and stimulate globulins that bind to sex hormones, which then binds estrogens that increase the risk to cancer. Lignans also inhibit the activity of aromatase enzymes, which in turn, decreases the level of estrogens hormones. Some new research about lignans is concentrated on finding mechanisms of how flaxseed affects hormone production and estrogen metabolites in urine that are linked to increasing in risk for developing cancer. When flaxseed is part of the diet, it has been demonstrated to reduce the concentration of estrone sulfate and 17-beta-estradiol in serum significantly. The increase of the 2:16 alpha-hydroxyestrone ratio and level of 2-hydroxyestrone excretion in urine were observed in studies of pre-menopausal and post-menopausal women with flaxseed supplementation. As flaxseed is high in omega-3 fatty acids and influences hormones production, this observation led to an investigation of its possible role in preventing the onset of prostate cancer. Phase II clinical study is underway to investigate the role of flaxseed in prostate cancer chemoprevention.

Other Promising Natural Agents

In addition to the natural compounds listed earlier, few other promising natural agents are being actively investigated for their effectiveness in chemoprevention. Several of these molecules have shown encouraging anticancer activities. Some of these molecules include polyunsaturated fatty acids, ginkolide B, ellagic acid, some triterpenes like lupeol, betulinic acid, ginsenosides and oleanolic acid. Ellagic acid is an antioxidant polyphenol found in many fruits and vegetables including strawberries, grapes, pomegranates, raspberries, and nuts. It exhibits chemopreventive property against lung, skin, colon, bladder, colon, prostate, esophageal and breast cancers. Among the triterpenes, betulinic acid and lupeol have been studied in detail for their chemopreventive properties. They show a wide range of activity against multiple cancer types in both animal models as well as cell culture models. Among the studied polyunsaturated fatty acids, linoleic acid and its derivatives show promising results. These exhibited chemopreventive activity in animal models of breast, prostate and colon cancers. Currently, many preventive studies are being carried out for the efficacy of these agents in preventing cancers.

MOLECULAR TARGETS FOR NATURAL CHEMOPREVENTIVE AGENTS

Many of the cell signaling pathways are activated by natural compounds present in the diet. Different pathways are targeted by different chemoprotective agents. Moreover, the same compound sometimes activates different signaling pathways in different cell types. Some of these pathways and molecular targets are described now.

p53 Family Members

p53 has many mechanisms to show its anticancer property and have a role in genomic stability, apoptosis, and inhibiting angiogenesis [51, 52]. P53 activates DNA repair proteins when DNA has sustained damage. After its activation, p53 binds to regulatory DNA sequences and activates the expression of some genes. These target genes can be grouped into four groups- cell cycle inhibition, apoptosis, genetic stability and inhibition of angiogenesis [53]. P53 can function both as transactivation as well as transrepressor. Because of these roles in DNA protection, p53 has been considered as an important molecule for maintaining the integrity of the genome. Many natural chemopreventive agents have been shown to induce cell cycle arrest or apoptosis by activating p53 and its target genes. Curcumin has been shown to induce apoptosis through p53-dependent BAX induction in human bladder cancer and breast cancer cells. Curcumin also caused the p53-mediated apoptosis by activating its mitochondrial translocation. The expression of p53 and its target p21 and BAX were induced by EGCG in prostate cancer cells with wild-type p53. This activity was not observed in cancer deficient in p53. Both, p53 and BAX were also activated by EGCG in breast cancer cells. EGCG induces apoptosis by activating p73-dependent expression of a subset of p53 target genes including p21, reprimo, G1, PERP, MDM2, WIG1, and PIG11. Luteolin induced apoptosis and cell cycle arrest and increased chemosensitization by activating p53 and its target p21, PUMA, and BAX. p73 is also activated in response to EGCG in multiple myeloma cells. Genistein induced G2/M arrest and apoptosis in human malignant glioma cell lines by activating p53 and p21. The expression of p21, p27, PUMA, BAX, MDM2, and cyclin G is activated by resveratrol. All these genes are downstream targets of p53. Many other natural molecules have been reported to induce apoptosis and cell cycle arrest in cells that lack a functional copy of the p53 gene.

Activator Protein 1

Activator protein 1 (AP-1) is a heterodimeric basic region leucine zipper transcription factor. It is composed of proteins belonging to the c-Fos, c-Jun, ATF and JDP families and binds either to AP-1 DNA recognition elements or to cAMP response elements and leads to activation of their target genes. It regulates gene expression to a variety of stimuli, including growth factors, stress cytokines, and bacterial and viral infections. AP-1, in turn, controls a number of cellular processes including differentiation, proliferation, and apoptosis. In addition to being transcriptional activators, some evidence suggests that some of the biologic effects of AP-1 are mediated by repression of other genes. Some of the genes that are modulated by AP-1 include angiogenesis, differentiation, modulators of invasion and metastasis, proliferation, and survival [54]. Many natural chemopreventive agents like green tea [55], resveratrol [56], and curcumin [57] have been demonstrated to their ability for suppressing AP-1 activation and modulating AP-1 target genes. This suppression has been implicated to their effectiveness of chemopreventive properties. The transcriptional activity of AP-1 is inhibited by polyphenols present in green tea. This inhibition is an essential aspect of their growth inhibition of cancer cells. It has been shown that administration of resveratrol inhibited TPA-induced AP-1 DNA binding by inhibiting the nuclear expression of c-Jun and c-Fos. Some studies point that proapoptotic function of curcumin is mediated by the suppression of Ap-1. Curcumin also reduces cell survival of human glioma cells. This activity of curcumin can be correlated with the inhibition of AP-1 and NF-κB signaling pathways.

Nuclear Factor-Kappa B

NF-κB (nuclear factor kappa-light-chain-enhancer of activated B cells) is a crucial transcription factor. It plays a role in a number of physiological and pathological conditions. It exerts its activity by binding within promoters/enhancers of target genes. It then recruits coactivators and corepressors [58]. The transcription factor family of NF-κB consists of five members, p50, p52, p65, c-Rel, and Rel B. All the members share an N-terminal Rel homology domain that is responsible for

DNA binding and homo- and heterodimerization. NF-κB is activated by inflammatory stimuli, free radicals, cytokines, tumor promoters, carcinogens, endotoxins, ultraviolet light, γ-radiation, and X-rays. In turn, NF-κB target genes are induced that are important for suppression of apoptosis, cellular growth, and transformation, invasion, chemo resistance, radio resistance, metastasis and inflammation. Most of the natural chemopreventive agents, including resveratrol, curcumin, lycopene, EGCG, luteolin, and genistein act as strong inhibitors of NF-κB pathways [59, 60]. These compounds may block one or more steps in the NF-κB signaling pathway. Examples include translocation of NF-κB to the nucleus, inhibition of the most upstream growth factor receptors that activate the NF-κB signaling cascade, interactions with the basal transcriptional machinery or DNA binding of the dimers.

Growth Factors and Their Receptors

Growth factors are signaling molecules that bind to receptors on the cell surface and are capable of stimulating cellular growth, proliferation, and cellular differentiation. Many growth factors, such as platelet-derived growth factor, endothelial growth factor, fibroblast growth factor, insulin-like growth factor, transforming growth factor, and colony-stimulating growth factor are linked to cancer. Abnormal growth factor signaling pathways lead to suppression of apoptotic signals, increased cell proliferation and invasion that contribute to metastasis. After growth factors activate the receptor, several signaling pathways downstream in the activation pathway are activated. Most important of these pathways are PI3K-AKT and Ras-MAPK. These signaling pathways are very important in cancer development and these become target for several chemopreventive molecules and molecules used for the treatment of cancer. Curcumin inhibits the ligand-stimulated activation of EGFR and enhances the growth inhibitory effects of FU and oxaliplatin. Both these activities are mediated through insulin-like growth factor receptor and EGFR pathways. When curcumin was administered with FOLPOX, decreased expression and activation of HER-2, HER-3, EGFR, and IGF-1R and their downstream effectors such as COX-2 and AKT were reported. Several studies have demonstrated that chemopreventive activity of GTP depends on inhibition of growth factor signaling. Treatment of patients with head and neck cancer with EGCG inhibited EGFR phosphorylation and its downstream targets AKT and ERK. It also potentiates the activity of the erlotinib, a tyrosine kinase inhibitor. Ubiquitin-mediated degradation of EGFR and its internalization was mediated by EGCG and ultimately undermined EGFR signaling. There are some studies that suggest the inhibitory effect of EGCG on activation of EGFR was associated with altered lipid order in HT29 colon cancer cells. EGCG also inhibited the activation of IGF-1 receptor present on the human colon cancer cells. Luteolin inhibited MAPK/ERK signaling and IGF-1–induced activation. Luteolin mediated apoptosis and growth inhibition of pancreatic tumor cells was associated with the inhibition of tyrosine kinase activity of EGFR.

Signal Transducers and Transcriptional Activators

Investigation of interferon-mediated transcriptional activation resulted in the identification of a previously unknown nuclear signal transduction pathway. Many tyrosine kinases, when activated, lead to phosphorylation, dimerization, and nuclear localization of the signal transducers and activators of transcription (STAT) proteins. These proteins control the transcription after binding to specific DNA elements. Seven mammalian STAT family members (STAT1, STAT2, STAT3, STAT4, STAT5A, STAT5B, and STAT6) have been studied in detail and recombinant expressed. They all share common structural motifs. Constitutive activation of STAT3 and STAT5 has been implicated in multiple lymphomas, myelomas, leukemias, and several solid tumors [61]. STAT activation in cancer cells is modulated by several natural chemopreventive agents like green tea, resveratrol, and curcumin. Oral administration of green tea polyphenols significantly reduced the expression of STAT3 in mice with prostate cancer. This activity may be linked to apoptosis and growth inhibition and this mice model. It has been observed that resveratrol modulates the expression of interleukin---induced ICAM-1 gene by changing phosphorylation of the STAT3 protein. Several studies have also suggested that role of STAT signaling pathways in curcumin-mediated chemoprevention. Curcumin prevented the tumor-induced apoptosis of T-cells by inducing expression of Bcl-2. It has been demonstrated that luteolin reduces the phosphorylation level of STAT3, which in turn, is targeted for proteasomal degradation. Luteolin also reduces the expression level of many proteins related to cancer like, VEGF, cyclin D1, Bcl-x(L), and survivin.

Immunoprevention

Immunoprevention is an approach to prevent cancer by stimulating the host immune system to destroy damaged cells before cancer develops. It is well known now that the one is more likely to get cancer in the absence of functional T cells or cytokines derived from T cells. So, one approach to cancer prevention may be to activate the functional T cells or produce increase the production of certain cytokines. For *e.g.*, INF-γ and IL-12 cytokines may be used to prevent cancer. Many natural molecules have been identified that are capable of modifying host factors. These may be implications in their cancer chemoprevention properties. In one study, it has been reported that GTP induces IL-12 dependent DNA repair, activates cytotoxic (CD8+) T cells, induces apoptosis of tumor cells and inhibits angiogenic factors. This activity ultimately leads to the inhibition of skin cancer. EGCG increased the $CD8^+$ T-cell mediated antitumor immunity induced by DNA vaccination [61]. Upon oral administration of curcumin to mouse models of cancer, progenitor, effector and circulating T-cells were regenerated. The administration of curcumin to tumor-bearing animals resulted in the restoration of the progenitor, effector, and circulating T cells. In another study, genistein was demonstrated to modulate the host immune response. Host became more resistant to the tumor and this can be correlated to the increase in the activity of natural killer cells and cytotoxic T cells. Many host reactions contribute to the development of cancers, including stroma, chronic inflammation, an expanding vasculature, in addition to tumor cells themselves. This suggests that these host factors could be targeted in order to delay cancer progression as well as prevent from originating. Indeed, a few natural molecules have been identified that target these factors. When xenograft mouse was administered luteolin, it inhibited VEGF-dependent angiogenesis and growth of the tumor. A number of studies suggest that chemoprevention by curcumin is mediated by targeting angiogenesis.

CONCLUDING REMARKS

Chemoprevention is becoming an increasingly popular research area to study the methods to prevent the cancer onset. USFDA approval of the use of tamoxifen and raloxifene for breast cancer risk reduction helped draw attention toward chemoprevention. Several epidemiological and preclinical findings, as well as, several clinical studies strongly suggest the important role of natural products present in the diet in preventing and treating various cancers. Multiple signaling transduction pathways are targeted by natural molecules. These pathways vary widely according to the type of cancer. The important question to the researcher community is how to use all this available information in preventing cancer from developing in a population that has a different risk for various types of cancers. Another aspect of this problem is that many of these natural molecules are not available in large amounts in dietary products and their potency is low. The chemical synthesis of analogs of natural compounds may be a solution for these limitations related to potency and bioavailability. As an example, the synthetic analog of curcumin, EF24, was found to be more than ten times more potent than its natural version [62]. A few natural molecules have been identified that show synergistic effect with known agents for chemoprevention or with other natural molecules. The toxicity associated with synthetic drugs still remain a significant barrier for currently available chemopreventive and chemotherapeutic drugs. Using natural compounds with current chemotherapeutic agents may help to minimize toxicity associated with synthetic drugs as these are safer, as compared to synthetic molecules. As an example, genistein was demonstrated to sensitize prostate cancer to radiation in animal studies. A clinical trial suggested that soy isoflavones could be helpful in preventing adverse effects to bladder and bowl caused by radiation. A better understanding of carcinogenesis, new technologies about screening and detecting cancer at early stages, identification of new targets are leading the way to early-phase clinical trials. With the advancement of research for natural chemopreventive molecules, new avenues would be available for preventing or treating the various cancer types in the human population. However, more preclinical studies and clinical trials are required to prove the usefulness of these natural compounds in chemoprevention.

The treatment of cancer has been significantly improved since the beginning of modern research on cancer. Many human cancers, like lymphomas and testicular cancer, can be completely cured now with available medical intervention. In some other cases, the survival of cancer patients can be enhanced. The research towards identification and purification of natural molecules have played an important role towards this progress. Nature is an important source of these molecules. Several of these natural products are being used by clinicians around the world in preventing and curing cancer. These natural compounds are either cytotoxic as they target non-specific biomolecules that are expressed by cancer cells or to a lesser extent by healthy dividing cells or they target macromolecules that are expressed only by the cancer cells. An important point to note is that mere cytotoxic activity of a molecule is not sufficient to treat it as an antitumor molecule. Further experiments, as well as, clinical trials are needed to show the importance of these molecules in chemoprevention and the treatment of cancer.

Although a good number of studies support using natural agents for chemoprevention, preclinical and clinical data in this field are still not sufficient. In this regard, it is important to realize that a natural molecule to be qualified and certified to be used as a chemopreventive agent, extensive preclinical and clinical studies, as well as, documentation of its pharmacological properties are required. Cancer causes a lot of emotional and psychological stress, financial burden and physical pain. Naturally, cancer patients and their families desperately search for the treatments for cancer that their physicians may be unable to provide. A controversial aspect is the use of natural compounds, nutraceuticals, and other complementary approaches, that may not be scientifically proven safe or effective in the treatment of cancer. The media attention, in this case, plays an important role. Sometimes, the expectation from these alternative therapies for treating cancer are not realistic and difficult to achieve. Therefore, taking cautious approach is advisable until these alternative therapies are proven safe and effective. The use of complementary therapies in treating cancer, in addition to standard chemotherapy may not always give the desired results.

CONFLICT OF INTEREST

The authors confirm that they have no conflict of interest to declare for this publication.

ACKNOWLEDGEMENTS

Declared none.

REFERENCES

[1]   Jemal A, Siegel R, Ward E, *et al.* Cancer statistics, 2008. CA Cancer J Clin 2008; 58(2): 71-96.
        [http://dx.doi.org/10.3322/CA.2007.0010] [PMID: 18287387]

[2]   World cancer research fund/ American institute for cancer research. Food, Nutrition, Physical activity, and the Prevention of Cancer: a Global Perspective. Washington DC: AICR 2007.

[3]   NCID. Clinical development plan: curcumin. J Cell Biochem Suppl 1996; 26(Suppl): 72-85.
        [PMID: 9154169]

[4]   Rao CV, Rivenson A, Simi B, Reddy BS. Chemoprevention of colon carcinogenesis by dietary curcumin, a naturally occurring plant phenolic compound. Cancer Res 1995; 55(2): 259-66.
        [PMID: 7812955]

[5]   Khan N, Afaq F, Mukhtar H. Cancer chemoprevention through dietary antioxidants: progress and promise. Antioxid Redox Signal 2008; 10(3): 475-510.
        [http://dx.doi.org/10.1089/ars.2007.1740] [PMID: 18154485]

[6]   Li M, Zhang Z, Hill DL, Wang H, Zhang R. Curcumin, a dietary component, has anticancer, chemosensitization, and radiosensitization effects by down-regulating the MDM2 oncogene through the PI3K/mTOR/ETS2 pathway. Cancer Res 2007; 67(5): 1988-96.
        [http://dx.doi.org/10.1158/0008-5472.CAN-06-3066] [PMID: 17332326]

[7]   Kuttan G, Kumar KB, Guruvayoorappan C, Kuttan R. Antitumor, anti-invasion, and antimetastatic effects of curcumin. Adv Exp Med Biol 2007; 595: 173-84. [PubMed].
        [http://dx.doi.org/10.1007/978-0-387-46401-5_6] [PMID: 17569210]

[8]   Verma SP, Salamone E, Goldin B. Curcumin and genistein, plant natural products, show synergistic inhibitory effects on the growth of human breast cancer MCF-7 cells induced by estrogenic pesticides. Biochem Biophys Res Commun 1997; 233(3): 692-6.
        [http://dx.doi.org/10.1006/bbrc.1997.6527] [PMID: 9168916]

[9]   Khafif A, Schantz SP, Chou TC, Edelstein D, Sacks PG. Quantitation of chemopreventive synergism between (-)-epigallocatechin-3-gallate and curcumin in normal, premalignant and malignant human oral epithelial cells. Carcinogenesis 1998; 19(3): 419-24.
        [http://dx.doi.org/10.1093/carcin/19.3.419] [PMID: 9525275]

[10]   Sreepriya M, Bali G. Effects of administration of Embelin and Curcumin on lipid peroxidation, hepatic glutathione   antioxidant   defense   and   hematopoietic   system   during   N-nitrosodiethylamine/Phenobarbital-induced hepatocarcinogenesis in Wistar rats. Mol Cell Biochem 2006; 284(1-2): 49-55.
        [http://dx.doi.org/10.1007/s11010-005-9012-7] [PMID: 16477385]

[11]   Cheng AL, Hsu CH, Lin JK, *et al.* Phase I clinical trial of curcumin, a chemopreventive agent, in patients with high-risk or pre-malignant lesions. Anticancer Res 2001; 21(4B): 2895-900. [PMID: 11712783]

[12]   Baliga MS, Meleth S, Katiyar SK. Growth inhibitory and antimetastatic effect of green tea polyphenols on metastasis-specific mouse mammary carcinoma 4T1 cells *in vitro* and *in vivo* systems. Clin Cancer Res 2005; 11(5): 1918-27. [http://dx.doi.org/10.1158/1078-0432.CCR-04-1976] [PMID: 15756018]

[13]   Sartippour MR, Heber D, Ma J, Lu Q, Go VL, Nguyen M. Green tea and its catechins inhibit breast cancer xenografts. Nutr Cancer 2001; 40(2): 149-56. [http://dx.doi.org/10.1207/S15327914NC402_11] [PMID: 11962250]

[14]   Xiao H, Hao X, Simi B, *et al.* Green tea polyphenols inhibit colorectal aberrant crypt foci (ACF) formation and prevent oncogenic changes in dysplastic ACF in azoxymethane-treated F344 rats. Carcinogenesis 2008; 29(1): 113-9. [http://dx.doi.org/10.1093/carcin/bgm204] [PMID: 17893236]

[15]   Shanafelt TD, Lee YK, Call TG, *et al.* Clinical effects of oral green tea extracts in four patients with low grade B-cell malignancies. Leuk Res 2006; 30(6): 707-12. [http://dx.doi.org/10.1016/j.leukres.2005.10.020] [PMID: 16325256]

[16]   Pisters KM, Newman RA, Coldman B, *et al.* Phase I trial of oral green tea extract in adult patients with solid tumors. J Clin Oncol 2001; 19(6): 1830-8. [http://dx.doi.org/10.1200/JCO.2001.19.6.1830] [PMID: 11251015]

[17]   Jang M, Cai L, Udeani GO, *et al.* Cancer chemopreventive activity of resveratrol, a natural product derived from grapes. Science 1997; 275(5297): 218-20. [http://dx.doi.org/10.1126/science.275.5297.218] [PMID: 8985016]

[18]   Athar M, Back JH, Tang X, *et al.* Resveratrol: a review of preclinical studies for human cancer prevention. Toxicol Appl Pharmacol 2007; 224(3): 274-83. [http://dx.doi.org/10.1016/j.taap.2006.12.025] [PMID: 17306316]

[19]   Khan N, Afaq F, Mukhtar H. Cancer chemoprevention through dietary antioxidants: progress and promise. Antioxid Redox Signal 2008; 10(3): 475-510. [http://dx.doi.org/10.1089/ars.2007.1740] [PMID: 18154485]

[20]   Ma X, Tian X, Huang X, Yan F, Qiao D. Resveratrol-induced mitochondrial dysfunction and apoptosis are associated with Ca2+ and mCICR-mediated MPT activation in HepG2 cells. Mol Cell Biochem 2007; 302(1-2): 99-109. [http://dx.doi.org/10.1007/s11010-007-9431-8] [PMID: 17396234]

[21]   Kubota T, Uemura Y, Kobayashi M, Taguchi H. Combined effects of resveratrol and paclitaxel on lung cancer cells. Anticancer Res 2003; 23(5A): 4039-46. [PMID: 14666716]

[22]   Canene-Adams K, Lindshield BL, Wang S, Jeffery EH, Clinton SK, Erdman JW Jr. Combinations of tomato and broccoli enhance antitumor activity in dunning r3327-h prostate adenocarcinomas. Cancer Res 2007; 67(2): 836-43. [http://dx.doi.org/10.1158/0008-5472.CAN-06-3462] [PMID: 17213256]

[23]   Giovannucci E. Tomatoes, tomato-based products, lycopene, and cancer: review of the epidemiologic literature. J Natl Cancer Inst 1999; 91(4): 317-31. [http://dx.doi.org/10.1093/jnci/91.4.317] [PMID: 10050865]

[24]   García-Closas R, Castellsagué X, Bosch X, González CA. The role of diet and nutrition in cervical carcinogenesis: a review of recent evidence. Int J Cancer 2005; 117(4): 629-37. [http://dx.doi.org/10.1002/ijc.21193] [PMID: 15912536]

[25]   Schwarz S, Obermüller-Jevic UC, Hellmis E, Koch W, Jacobi G, Biesalski HK. Lycopene inhibits disease progression in patients with benign prostate hyperplasia. J Nutr 2008; 138(1): 49-53.
[PMID: 18156403]

[26]   Kim HS, Bowen P, Chen L, et al. Effects of tomato sauce consumption on apoptotic cell death in prostate benign hyperplasia and carcinoma. Nutr Cancer 2003; 47(1): 40-7.
[http://dx.doi.org/10.1207/s15327914nc4701_5] [PMID: 14769536]

[27]   Kucuk O, Sarkar FH, Sakr W, et al. Phase II randomized clinical trial of lycopene supplementation before radical prostatectomy. Cancer Epidemiol Biomarkers Prev 2001; 10(8): 861-8.
[PMID: 11489752]

[28]   Sahin K, Ozercan R, Onderci M, et al. Lycopene supplementation prevents the development of spontaneous smooth muscle tumors of the oviduct in Japanese quail. Nutr Cancer 2004; 50(2): 181-9.
[http://dx.doi.org/10.1207/s15327914nc5002_8] [PMID: 15623465]

[29]   Kim ND, Mehta R, Yu W, et al. Chemopreventive and adjuvant therapeutic potential of pomegranate (Punica granatum) for human breast cancer. Breast Cancer Res Treat 2002; 71(3): 203-17.
[http://dx.doi.org/10.1023/A:1014405730585] [PMID: 12002340]

[30]   Toi M, Bando H, Ramachandran C, et al. Preliminary studies on the anti-angiogenic potential of pomegranate fractions in vitro and in vivo. Angiogenesis 2003; 6(2): 121-8.
[http://dx.doi.org/10.1023/B:AGEN.0000011802.81320.e4] [PMID: 14739618]

[31]   Zhang Y, Vareed SK, Nair MG. Human tumor cell growth inhibition by nontoxic anthocyanidins, the pigments in fruits and vegetables. Life Sci 2005; 76(13): 1465-72.
[http://dx.doi.org/10.1016/j.lfs.2004.08.025] [PMID: 15680311]

[32]   Khan N, Hadi N, Afaq F, Syed DN, Kweon MH, Mukhtar H. Pomegranate fruit extract inhibits prosurvival pathways in human A549 lung carcinoma cells and tumor growth in athymic nude mice. Carcinogenesis 2007; 28(1): 163-73.
[http://dx.doi.org/10.1093/carcin/bgl145] [PMID: 16920736]

[33]   Khan N, Afaq F, Kweon MH, Kim K, Mukhtar H. Oral consumption of pomegranate fruit extract inhibits growth and progression of primary lung tumors in mice. Cancer Res 2007; 67(7): 3475-82.
[http://dx.doi.org/10.1158/0008-5472.CAN-06-3941] [PMID: 17389758]

[34]   Malik A, Afaq F, Sarfaraz S, Adhami VM, Syed DN, Mukhtar H. Pomegranate fruit juice for chemoprevention and chemotherapy of prostate cancer. Proc Natl Acad Sci USA 2005; 102(41): 14813-8.
[http://dx.doi.org/10.1073/pnas.0505870102] [PMID: 16192356]

[35]   Sartippour MR, Seeram NP, Rao JY, et al. Ellagitannin-rich pomegranate extract inhibits angiogenesis in prostate cancer in vitro and in vivo. Int J Oncol 2008; 32(2): 475-80.
[PMID: 18202771]

[36]   Pantuck AJ, Leppert JT, Zomorodian N, et al. Phase II study of pomegranate juice for men with rising prostate-specific antigen following surgery or radiation for prostate cancer. Clin Cancer Res 2006; 12(13): 4018-26.
[http://dx.doi.org/10.1158/1078-0432.CCR-05-2290] [PMID: 16818701]

[37]   Shimoi K, Saka N, Kaji K, Nozawa R, Kinae N. Metabolic fate of luteolin and its functional activity at focal site. Biofactors 2000; 12(1-4): 181-6.
[http://dx.doi.org/10.1002/biof.5520120129] [PMID: 11216484]

[38]   Yang SF, Yang WE, Chang HR, Chu SC, Hsieh YS. Luteolin induces apoptosis in oral squamous cancer cells. J Dent Res 2008; 87(4): 401-6.
[http://dx.doi.org/10.1177/154405910808700413] [PMID: 18362328]

[39]   Ju W, Wang X, Shi H, Chen W, Belinsky SA, Lin Y. A critical role of luteolin-induced reactive oxygen species in blockage of tumor necrosis factor-activated nuclear factor-kappaB pathway and sensitization of apoptosis in lung cancer cells. Mol Pharmacol 2007; 71(5): 1381-8.
[http://dx.doi.org/10.1124/mol.106.032185] [PMID: 17296806]

[40]   Zhang Q, Zhao XH, Wang ZJ. Flavones and flavonols exert cytotoxic effects on a human oesophageal adenocarcinoma cell line (OE33) by causing G2/M arrest and inducing apoptosis. Food Chem Toxicol 2008; 46(6): 2042-53.
[http://dx.doi.org/10.1016/j.fct.2008.01.049] [PMID: 18331776]

[41]   Selvendiran K, Koga H, Ueno T, *et al.* Luteolin promotes degradation in signal transducer and activator of transcription 3 in human hepatoma cells: an implication for the antitumor potential of flavonoids. Cancer Res 2006; 66(9): 4826-34.
[http://dx.doi.org/10.1158/0008-5472.CAN-05-4062] [PMID: 16651438]

[42]   Onozawa M, Kawamori T, Baba M, *et al.* Effects of a soybean isoflavone mixture on carcinogenesis in prostate and seminal vesicles of F344 rats. Jpn J Cancer Res 1999; 90(4): 393-8.
[http://dx.doi.org/10.1111/j.1349-7006.1999.tb00760.x] [PMID: 10363576]

[43]   Lakshman M, Xu L, Ananthanarayanan V, *et al.* Dietary genistein inhibits metastasis of human prostate cancer in mice. Cancer Res 2008; 68(6): 2024-32.
[http://dx.doi.org/10.1158/0008-5472.CAN-07-1246] [PMID: 18339885]

[44]   Hebert JR, Hurley TG, Olendzki BC, Teas J, Ma Y, Hampl JS. Nutritional and socioeconomic factors in relation to prostate cancer mortality: a cross-national study. J Natl Cancer Inst 1998; 90(21): 1637-47.
[http://dx.doi.org/10.1093/jnci/90.21.1637] [PMID: 9811313]

[45]   Ziegler RG, Hoover RN, Pike MC, *et al.* Migration patterns and breast cancer risk in Asian-American women. J Natl Cancer Inst 1993; 85(22): 1819-27.
[http://dx.doi.org/10.1093/jnci/85.22.1819] [PMID: 8230262]

[46]   Goodman MT, Wilkens LR, Hankin JH, Lyu LC, Wu AH, Kolonel LN. Association of soy and fiber consumption with the risk of endometrial cancer. Am J Epidemiol 1997; 146(4): 294-306.
[http://dx.doi.org/10.1093/oxfordjournals.aje.a009270] [PMID: 9270408]

[47]   Pendleton JM, Tan WW, Anai S, *et al.* Phase II trial of isoflavone in prostate-specific antigen recurrent prostate cancer after previous local therapy. BMC Cancer 2008; 8: 132.
[http://dx.doi.org/10.1186/1471-2407-8-132] [PMID: 18471323]

[48]   Pelley RP, Strickland FM. Plants, polysaccharides, and the treatment and prevention of neoplasia. Crit Rev Oncog 2000; 11(3-4): 189-225.
[http://dx.doi.org/10.1615/CritRevOncog.v11.i34.10] [PMID: 11358267]

[49]   Fisher M, Yang LX. Anticancer effects and mechanisms of polysaccharide-K (PSK): implications of cancer immunotherapy. Anticancer Res 2002; 22(3): 1737-54.
[PMID: 12168863]

[50]   Watson GW, Wickramasekara S, Palomera-Sanchez Z. SUV39H1/H3K9me3 attenuates sulforaphane-induced apoptotic signaling in PC3 prostate cancer cells. Oncogenesis. 2014; 3: p. (12)e131.
[http://dx.doi.org/10.1038/oncsis.2014.47] [PMCID: PMC4275561]

[51]   Hanel W, Moll UM. Links between mutant p53 and genomic instability. J Cell Biochem 2012; 113(2): 433-9.
[PMID: 22006292]

[52]   Bode AM, Dong Z. Post-translational modification of p53 in tumorigenesis. Nat Rev Cancer 2004; 4(10): 793-805.
[http://dx.doi.org/10.1038/nrc1455] [PMID: 15510160]

[53]   Carr AM. Cell cycle. Piecing together the p53 puzzle. Science 2000; 287(5459): 1765-6.
[http://dx.doi.org/10.1126/science.287.5459.1765] [PMID: 10755928]

[54]   Shaulian E, Karin M. AP-1 as a regulator of cell life and death. Nat Cell Biol 2002; 4(5): E131-6.
[http://dx.doi.org/10.1038/ncb0502-e131] [PMID: 11988758]

[55]   Kim HS, Kim MH, Jeong M, *et al.* EGCG blocks tumor promoter-induced MMP-9 expression *via* suppression of MAPK and AP-1 activation in human gastric AGS cells. Anticancer Res 2004; 24(2B): 747-53.
[PMID: 15161022]

[56]   Kundu JK, Shin YK, Surh YJ. Resveratrol modulates phorbol ester-induced pro-inflammatory signal transduction pathways in mouse skin *in vivo*: NF-kappaB and AP-1 as prime targets. Biochem Pharmacol 2006; 72(11): 1506-15.
[http://dx.doi.org/10.1016/j.bcp.2006.08.005] [PMID: 16999939]

[57]   Dhandapani KM, Mahesh VB, Brann DW. Curcumin suppresses growth and chemoresistance of human glioblastoma cells *via* AP-1 and NFkappaB transcription factors. J Neurochem 2007; 102(2): 522-38.

[58]   Aggarwal BB. Nuclear factor-kappaB: the enemy within. Cancer Cell 2004; 6(3): 203-8.
[http://dx.doi.org/10.1016/j.ccr.2004.09.003] [PMID: 15380510]

[59]   Kim GY, Kim JH, Ahn SC, *et al.* Lycopene suppresses the lipopolysaccharide-induced phenotypic and functional maturation of murine dendritic cells through inhibition of mitogen-activated protein kinases and nuclear factor-kappaB. Immunology 2004; 113(2): 203-11.
[http://dx.doi.org/10.1111/j.1365-2567.2004.01945.x] [PMID: 15379981]

[60]   Li Y, Kucuk O, Hussain M, Abrams J, Cher ML, Sarkar FH. Antitumor and antimetastatic activities of docetaxel are enhanced by genistein through regulation of osteoprotegerin/receptor activator of nuclear factor-kappaB (RANK)/RANK ligand/MMP-9 signaling in prostate cancer. Cancer Res 2006; 66(9): 4816-25.
[http://dx.doi.org/10.1158/0008-5472.CAN-05-3752] [PMID: 16651437]

[61]   Kang TH, Lee JH, Song CK, *et al.* Epigallocatechin-3-gallate enhances CD8+ T cell-mediated antitumor immunity induced by DNA vaccination. Cancer Res 2007; 67(2): 802-11.
[http://dx.doi.org/10.1158/0008-5472.CAN-06-2638] [PMID: 17234792]

[62]   Kasinski AL, Du Yuhong, Thomas SL, *et al.* Inhibition of IκB kinase-nuclear factor-κB signaling pathway by 3,5- Bis (2-flurobenzylidene) piperidin-4-one (EF24), a novel monoketone analog of curcumin. Mol Pharm 2008; 74(3): 654-61.
[http://dx.doi.org/10.1124/mol.108.046201]

CHAPTER 5

African Medicinal Plants: An Untapped Reservoir of Potential Anticancer Agents

Conrad V. Simoben[1,2] and **Fidele Ntie-Kang**[*, 1,2]

¹ Pharmaceutical Chemistry, Martin-Luther Universität Halle-Wittenberg, Wolfgang-Langenbeck-Str. 4, 06120 Halle (Saale), Germany

² Chemistry Department, University of Buea, P. O. Box 63 Buea, South West Region, Cameroon

Abstract: Despite continuing scientific and commercial interests in cancer research around drug discovery, both less developed and developed countries are still trapped in the grip of this deadly and dreadful disease. Naturally occurring compounds represent approximately 50% of the chemotherapeutic agents, which have so far been brought to the market for cancer treatment. Traditional preparations have been the major source of cancer treatment in Africa, with traditional healers making regular use of these plants for the treatment of cancer and other ailments, since the continent is endowed with a rich floral bio-diversity. Africa's medicinal plants are known to biosynthesize interesting chemical structures with promising biological activities Thus, natural products from the African continent hold a premise for drug discovery and it is expected that the next generation of drugs, including potential anti-cancer drugs or the scaffolds necessary for the synthesis of new anti-cancer drugs could be lodged in African plants. We present some promising natural products for the anticancer drug development from African flora.

Keywords: Africa, Cancer, Medicinal plants, Natural products.

INTRODUCTION

Cancer in Africa

Overall, about 847,000 new cancer cases and an estimated 550,000 cancer deaths occurred in Africa in the year 2012 alone [1]; these estimates were expected to double in a few decades because of the adoption of western lifestyles, *e.g.* smoking, poor diet and little physical exercise, along with reproductive factors in economically developing/transitioning countries [2 - 5].

---

* **Corresponding author Fidele Ntie-Kang:** Department of Pharmaceutical Chemistry (Sippl Group), Institute for Pharmacy, Martin-Luther-Universität Halle-Wittenberg, Wolfgang-Langenbeck-Str. 4, 06120 Halle (Saale), Germany; Tel/Fax: +4915217812791; E-mails: ntiekfidele@gmail.com, fidele.ntie-kang@pharmazie.uni-halle.de

Sahdeo Prasad & Amit Kumar Tyagi (Eds.)

In spite of this growing burden, cancer continues to receive low public health priority in Africa, mainly due to limited resources and other pressing public health problems, including communicable diseases, *e.g.* the acquired immune deficiency syndrome (AIDS)/human immunodeficiency virus (HIV) infection, malaria and tuberculosis. The three most commonly diagnosed cancers in Africa are lung, stomach and liver in men while breast, cervix uteri and lung are reported for women [2].

Current Clinical Management of Cancer

Cancer is largely feared because it is known to be difficult to cure. Currently there are three major ways of treating cancer: radiotherapy (the use of high-energy radiation to kill/destroy cancer cells), surgery (the removal of infected cells) and the use of cytotoxic drugs (the killing of cancer cells), also known as drug therapy or chemotherapy, each treatment method having significant limitations [6].

Chemotherapy remains one of the main methods of modern cancer treatment by humans [2, 7 - 9]. However, many of these cytotoxic drugs have the potential to be very harmful to the body as they are rarely designed to be "cancer cell-specific" in their modes of action. In addition, adverse and/or side effect(s) resulting from cancer chemotherapy, *e.g.* fatigue, loss of appetite, nausea, vomiting, bowel changes such as diarrhoea, constipation, hair loss, reduced levels of red and white blood cells and platelets, mouth ulcers or mouth infections and skin problems such as itchiness or extreme light sensitivity of the available drugs, cost factors (high cost of the available drugs and the long exposure/treatment period for the disease) also constitute a major limitation [6, 9, 10].

Africans and the Use of Natural Products for Cancer Chemotherapy

The use of plants by humans as medicinal agents (employed as decoctions, concoctions, steam baths, burning of ashes and smearing of liquids and gas upon the patients, among other practices) pre-dates recorded history [11]. The development of formularies and pharmacopoeia from ethno-medical plant use data provides an important solution for global health care, as well as significantly contributing to drug development [12]. It is estimated that 60 – 85% of the world's population especially people living in the rural areas depend directly on plants as medicines for the treatment of various diseases [13 - 16]. Two main reasons lie behind this observation; the low purchasing power within the populations living in the rural areas and the limited access to modern drugs. It is common knowledge that medicinal plants, along with their phytochemicals, have been widely used for their curative purposes. They could therefore be considered as playing a paramount role as an important source of lead compounds for drug discovery for a variety of diseases [14, 17]. The continent of Africa is very rich in floral biodiversity and the use of these indigenous plants plays an important role in the treatment of several diseases [18, 19]. It is believed that the plant secondary metabolites hold enormous potential for drug discovery [20 - 23]. Reviews that empirically analyze bioactivity, ethnobotanical and ethnopharmacological uses of African plants have also been recently published [18, 24 - 27]. A lot of research efforts on the anticancer activities of plants in Africa used by traditional healers to treat and cure cancer related symptoms have been documented [10, 28 - 31].

SOME ISOLATED COMPOUNDS FROM AFRICAN MEDICINAL PLANTS WITH MEASURED ANTICANCER ACTIVITY

In Africa, traditional healers operate close to the populations, by making use of the biodiversity of plant species in such areas to cure various diseases and ailments. Several steps have been taken towards documenting the ethnobotanical and ethnopharmacological knowledge derived from the use of medicinal plants in Africa. The bioactivities of the metabolites have been recorded [15, 16, 26, 30 - 32]. The anti-cancer properties of some of the compounds identified in the current study have been previously described in some of our reviews [18, 24, 33 - 35]. The following sub-paragraphs discuss some of the isolated metabolites from African flora, along with the most promising compounds and compound classes (summarised in Table **1**).

Table 1. Some selected bioactive compounds from some African medicinal plants with promising anticancer activities.

Isolated metabolites	Plant species (Family)	Part of plant studied	Author, Reference
1-5	*Erythrina abyssinica* (Fabaceae)	Whole plant	Mohammed *et al.* [46]
6-8	*Erythrina excelsa* and *Erythrina senegalensis* (Fabaceae)	Roots	Kuete *et al.* [49]
9-15	*Ardisia kivuensis* (Myrsinaceae)	Stem	Ndontsa *et al.* [50, 51]
		Leaves and stem	
		Leaves and stem bark	Paul *et al.* [52]
16-18	*Albizia grandibracteata* (Fabaceae)	Stem bark	Krief *et al.* [53]
19-22	*Annona senegalensis* (Annonaceae)	Stem bark	Fatope and Audu [54]
23-28	*Allanblackia monticola* (Clusiaceae)	Seeds	Azebaze *et al.* [60]
29-35	*Hypericum riparium* (Guttiferae)	Roots	Tala *et al.* [62]
36 – 41	*Antiaris africana* (Moraceae)	Stem bark	Kuete *et al.* [63]

(Table 1) contd.....

Isolated metabolites	Plant species (Family)	Part of plant studied	Author, Reference
42	*Araliopsis synopsis* (Rutaceae)	Stem bark	Happi *et al.* [66]
43 and 44	*Hypericum lanceolatum* (Clusiaceae)	Leaves	Kuete *et al.* [68]
45 and 46	*Garcinia punctate* (Clusiaceae)		
47 and 48	*Cajanus cajan* (Fabaceae)	Leaves	Ashidi *et al.* [31]
49 - 53	*Elephantopus mollis* (Asteraceae)	Root and stem	Tabopda *et al.* [71]

Alkaloids

Alkaloids represent a group of naturally occurring nitrogen-containing organic compounds and are characterized by a bitter taste, which helps to protect plants from the aggression of animals [36]. A number of alkaloids isolated from African medicinal plants have exhibited potential anti-cancer activities (Fig. **1**) below. An example of such a plant is *Erythrina abyssinica* (Fabaceae, one of the largest plant families containing alkaloids) [37]. Plants in this family are known to be rich sources of bioactive alkaloids [38, 39] and flavonoids [40]. The isolated compounds have been found to display a variety of biological properties, such as antimicrobial [41, 42], anti-HIV-1 [43], anti-inflammatory [44] and anti-plasmodial activities [45].

Erythraline (1)   Erysodine (2)   Erysotrine (3)   8-Oxoerythraline (4)   11-Methoxyerysodine (5)

Fig. (1). Some alkaloids isolated from *Erythrina abyssinica*.

The *in vitro* cytotoxicity of the crude alkaloidal fraction of *E. abyssinica* against several anticancer cell lines (including HeLa, Hep-G2, HEP-2, HCT116, MCF-7 and HFB4) was investigated by Mohammed *et al.* This study showed that the plant extract had promising activity. IC_{50} values of 13.8, 10.1, 8.16, 13.9, 11.4 and 12.2 µg/mL respectively, were exhibited. Meanwhile 10 mg of the reference drug (Doxorubicin), used as a positive control, showed *in vitro* cytotoxicity with IC_{50} values of 3.64, 4.57, 4.89, 3.74, 2.97 and 3.96 mg/mL respectively [46]. Bioassay-guided fractionation and isolation of the crude alkaloidal fraction led to the isolation of five *Erythrina* alkaloids, identified as erythraline (**1**), erysodine (**2**), erysotrine (**3**), 8-oxoerythraline (**4**) and 11-methoxyerysodine (**5**). These compounds displayed *in vitro* cytotoxic activity against Hep-G2, with IC_{50} values of 17.60, 11.80, 15.80, 3.89 and 11.40 μg/mL respectively (Doxorubicin IC_{50} = 4.57 mg/mL) [46]. Furthermore, *in vitro* cytotoxic activity against HEP-2 was evaluated, resulting in IC_{50} values of 15.90, 19.90, 21.60, 18.50, 11.50 μg/mL respectively [46].

The authors further underwent a study of the structure-activity relationship (SAR) of the isolated alkaloids (**1 – 5**), which revealed that; for the HEP-2 cell line tested, the methylenedioxy group was an optimal substituent for compounds **1** to **3**, since neither a phenolic hydrogen bond donor in compound **2** nor the dimethoxy substituent in compound **3** could improve the HEP-2 cytotoxic activity of the alkaloids. Furthermore, the authors observed that the carbonyl group vicinal to the nitrogen atom is detrimental to the HEP-2 cytotoxic activity [46]. A contrary observation was carried out for the activity against Hep-G2 cell line. It was rather shown that compound **4** (having a keto group vicinal to the nitrogen atom), showed the highest cytotoxic potency against Hep-G2 cell lines. In addition, compound **3** (an unconstrained analogue of compound **1**) showed better cytotoxic potency, while the phenolic hydrogen bond donor in compound **2** confers to it a better potency than its corresponding methylated analogues (compounds **1** and **3**). It could therefore be concluded that the presence of α,β-unsaturated carbonyl group in 8-oxoerythraline could basically explain its highest cytotoxicity against Hep-G2 (with IC_{50} of 3.89 μg/mL), an activity which is even better than the reference drug Doxorubicin (with IC_{50} of 4.57 μg/mL).

Flavonoids

Flavonoids are polyphenolic molecules containing a basic skeleton with 15 carbon atoms and are soluble in water. They consist of two benzene rings connected by a short three carbon chain. Studies have shown that plant-derived flavonoids have displayed cytotoxic activities (Fig. **2** and Table **1**) against several cancer cell lines [47, 48]. Compounds **6** to **8** are amongst the most promising isolated flavonoids with anticancer activities. Neobavaisoflavone (**6**), sigmoidin H (**7**) and isoneorautenol (**8**) were derived from the methanol-dichloromethane (1:1) extracts of the roots of *Erythrina excelsa* and *Erythrina senegalensis* (Fabaceae) [49].

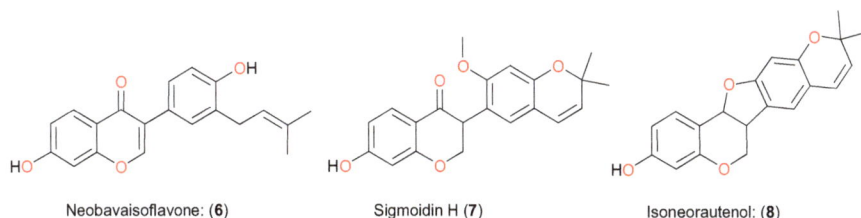

Neobavaisoflavone: (**6**)   Sigmoidin H (**7**)   Isoneorautenol: (**8**)

Fig. (2). Some cytotoxic flavonoids isolated from *Erythrina excelsa* and *Erythrina senegalensis*.

Compound **8** showed significant cytotoxicity toward drug sensitive and drug-resistant cancer cell lines, while compounds **6** and **7** were selectively active, with IC_{50} values ranging from 42.93 µM (toward CCRF-CEM cells) to 114.64 µM [against HCT116 ($p53^{+/+}$) cells] for compound **6** and 25.59 µM (toward U87MG) to 110.51 µM [against HCT116 ($p53^{+/+}$) cells] for compound **7**. IC_{50} values ranging from 2.67 µM (against MDA-MB 237BCRP cells) to 21.84 µM (toward U87MG) were measured for compound **8** as well as between 0.20 µM (toward CCRF-CEM cells) and 195.12 µM (toward CEM/ADR5000 cells) for Doxorubicin as the control drug.

Quinones

Quinones represent any member of a class of cyclic organic compounds containing two carbonyl groups, either adjacent or separated by a vinylene group, −CH = CH−, in a six-membered unsaturated ring. Some of the bioactive quinones with anti-cancer activities from African flora have been shown in Fig. (**3**) and Table **1**. These include ardisiaquinones J − P (Compounds **9-15**), which were isolated from *Ardisia kivuensis* (Myrsinaceae), a plant harvested from Mount Oku, Cameroon [50, 51]. Ardisiaquinones J (**9**) and K (**10**) were isolated from the methanol stem extract [50]. These compounds have shown considerable antiproliferative activity against several cell lines, *e.g.* Ishikawa, HeLa, MCF7 and A431, with IC_{50} values between 6.64 and 15.40 µM. In addition, ardisiaquinone K (**10**) showed a moderate free radical scavenging effect on DPPH-assay. In a further study, the aforementioned authors reported the isolation of ardisiaquinones L − P (compounds **11 − 15**) from the methanol extract of leaves and stems of the same plant, harvested around the same period and locality [51]. Compounds **11 − 15** showed cytotoxicity against *Artemia salina*, with the most active compounds being ardisiaquinone N (**13**) (mortality, 72%), ardisiaquinone M (**12**) (mortality 67%) and the mixture of ardisiaquinones N (**13**) and O (**14**) (mortality, 60%) at a concentration of 10 µg/mL, the compounds also demonstrating moderate antimicrobial activity [50, 51].

Additionally, Paul *et al.* identified the ardisiaquinones J (**9**), K (**10**), N (**13**) and a mixture of ardisiaquinones K (**10**) and P (**15**) from the stem bark and leaves of *A. kivuensis* and tested the isolates against Leukemia cell lines [52]. The report showed that compounds **9, 10, 13** and **15**, exhibit remarkable antiproliferative activity against the leukemia cell line TPH-1 with IC_{50} inhibition values of 2 to 2.1 µg/mL compared to 4 µg/mL for Paclitaxel used as the positive control.

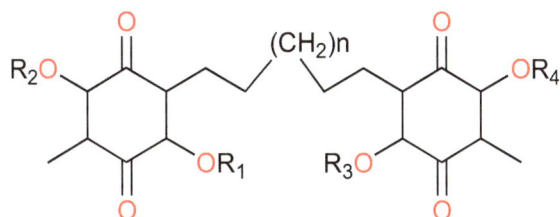

Ardisiaquinone J $R_1 = R_4 = H$; $R_2 = R_3 = Me$; n = 10 (**9**)
Ardisiaquinone K $R_1 = R_3 = R_4 = H$; $R_2 = Me$; n = 10 (**10**)
Ardisiaquinone L $R_1 = R_4 = H$; $R_2 = R_3 = Me$; n = 9(**11**)
Ardisiaquinone M $R_1 = R_3 = H$; $R_2 = R_4 = Me$; n = 9 (**12**)
Ardisiaquinone N $R_1 = R_3 = R_4 = H$; $R_2 = Me$; n = 9 (**13**)
Ardisiaquinone O $R_1 = Me$; $R_2 = R_3 = R_4 = H$; n = 9 (**14**)
Ardisiaquinone P $R_1 = Me$; $R_2 = R_3 = R_4 = H$; n = 10 (**15**)

Fig. (3). Some cytotoxic quinones isolated from *Ardisia kivuensis*.

Diterpenoids and Triterpenoids

Terpenoids represent a broad class of compounds, which are characterized by isoprene units and are well known to have remarkable activities against several diseases. Some of the most interesting bioactive diterpenoids and triterpenoids derived from African flora with promising anti-cancer properties are also shown in Table **1**, while their structures are shown in Fig. (**4**). The stem bark of *Albizia grandibracteata* (Fabaceae) is used in Uganda for the treatment of meteorism. Phytochemical investigation of the methanolic extract of the leaves of *A. grandibracteata*, carried out by Krief *et al*., yielded three novel oleanane-type triterpene saponins, namely grandibracteosides A (**16**), B (**17**) and C (**18**) [53]. The study reported that the crude extract, together with these compounds, exhibited significant inhibitory activity against two tumor cell lines, with IC_{50} values for compounds **16 - 18** of 0.04, 0.06 and 3.30 μM (KB cell line) and 0.008, 0.010 and 1.800 μM (MCF-7 cell line) respectively.

The bioassay-guided fractionation of *Annona senegalensis* (Annonaceae) stem bark by Fatope *et al*. gave four bioactive *ent*-kaurenoid diterpenoids, which include ent-3β-hydroxykaur-16-ene (**19**), *ent*-kaur-16-en-19-oic acid (**20**), ent-16,17-diacetoxykauran-16-oic acid (**21**) and *ent*-19-carbomethoxykauran-19-oic acid (**22**) [54]. Compound **20** showed selective and significant cytotoxicity against a broad range of cancer cell lines, including MCF-7 (breast cancer) cells (ED_{50} 1.0 *μ*g/mL), while **21** and **22** exhibited cytotoxic selectivity for PC-3 (prostate cancer) cells but with weaker potencies (ED_{50} 17-18 μg/mL).

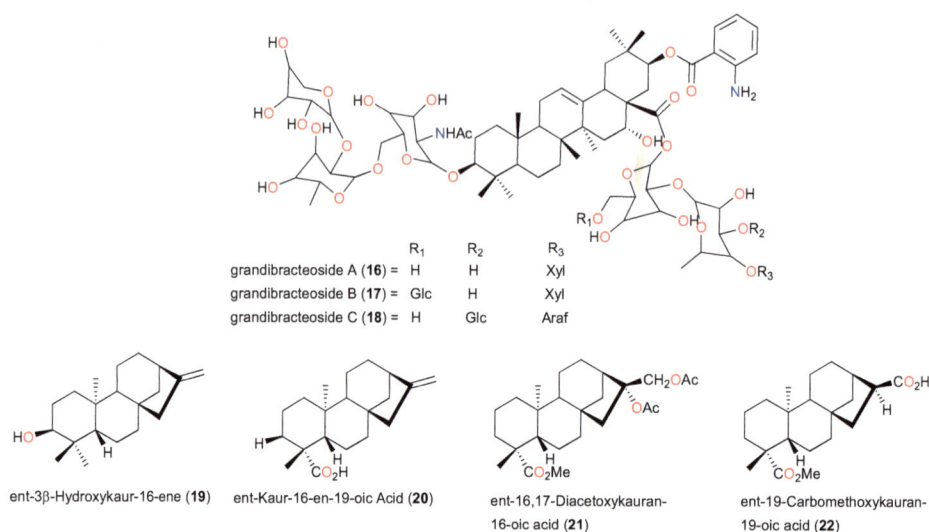

	R₁	R₂	R₃
grandibracteoside A (**16**) =	H	H	Xyl
grandibracteoside B (**17**) =	Glc	H	Xyl
grandibracteoside C (**18**) =	H	Glc	Araf

ent-3β-Hydroxykaur-16-ene (**19**)   ent-Kaur-16-en-19-oic Acid (**20**)   ent-16,17-Diacetoxykauran-16-oic acid (**21**)   ent-19-Carbomethoxykauran-19-oic acid (**22**)

Fig. (4). Some cytotoxic diterpenoids and triterpenoids isolated from *A. grandibracteata* (16 - 18) and *A. senegalensis* (19 - 22).

Xanthones

Xanthones possess the same structural backbone. What makes the various xanthones unique, are the side chains bound to the carbon atoms of the benzene rings at diverse positions. Xanthones are well known for their antitumor and chemo-preventive activity [55 - 59]. Test results for some of the most interesting bioactive xanthones derived from African flora with promising anti-cancer properties are as well summarized in Table **1** with their structures in Fig. (**5**).

The xanthones **23 – 28** were derived from the methylene chloride-methanol (1:1) extract from *Allanblackia monticola* (Clusiaceae) seeds. This plant has been used for the treatment of certain human ailments such as respiratory infections, diarrhoea and toothache. The compounds isolated have demonstrated apoptotic and antiproliferative activities against human leukemic B lymphocytes [60, 61], *e.g.* 1,7-dihydroxy-3-methoxy-2-(3-methylbut-2-enyl)xanthone (compound **23**), showed selective toxicity and proapoptotic activity on tumor cells from B-CLL patients [60].

Tala *et al.* also reported the isolation of compounds **29 – 35** for the first time, from the EtOAc-soluble portion of the methanol-dichloromethane (1:1) extract of the roots of *Hypericum riparium* (Guttiferae) [62]. While the extracts and two of the isolated compounds, cadensin D (**33**) and 5-hydroxy-1,3-dimethoxyxanthone (**34**), exhibited both antibacterial and antifungal activities that varied between the microbial species (MIC = 0.97 – 250 µg/mL), hypercalin C (**31**) showed potent cytotoxicity with LD_{50} of 3.23 µg/mL, confirming the plant's usage in African traditional medicine for the treatment of cancer.

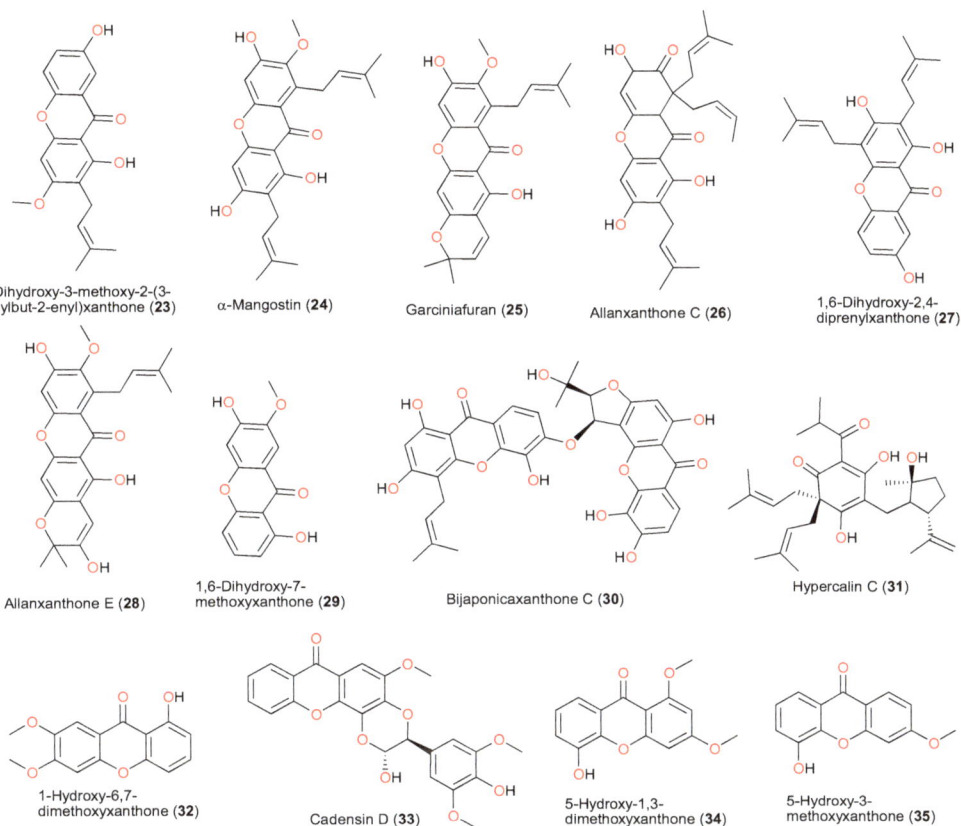

Fig. (5). Some cytotoxic xanthones isolated from *A. monticola* (23 - 28) and *H. riparium* (29 - 35).

Other Compound Classes

Other bioactive compounds isolated from the African flora with promising anti-cancer activities are shown in Figs. (**6 - 9**), also summarized in Table **1**. Among these compounds are the limonoids **36 – 40** and ellagic acid derivative (**41**), isolated from the methanol extract of *Antiaris africana* (Moraceae). This tree has been locally used in some Africa regions as timber, while its water and ethanol bark extracts are often used in traditional medicine for the treatment of chest pains and the leaf decoctions are applied in the treatment of syphilis [63 - 65]. Both the methanol extract and the isolated compound, 3,3'-dimethoxy-4'-*O*-β-*D*-xylopyronosylellagic acid (**41**), were shown to exhibit very high antioxidant activity (93.24 and 94.67%, respectively), which was not significantly different (p < 0.05) from that of Vitamin C (92.15%), the reference antioxidant drug employed in the study. This information supports the traditional use of *A. africana* for the treatment of cancer locally in Cameroon [63].

The coumarin scopoletin (**42**) was isolated from the dichloromethane extract of *Araliopsis synopsis* (Rutaceae) stem bark by Happi *et al.* [66]. This plant has been used traditionally for the treatment of sexually transmitted diseases [67]. Scopoletin (**42**) and five other compounds from this plant exhibited high cytotoxic activity against the human Caucasian prostate adenocarcinoma cell PC-3 line, with IC_{50} values ranging from 8.5 to 12.5 μM, when compared with the standard Doxorubicin ($IC_{50} = 0.9$ μM) [66].

Kuete*et al.* evaluated the compounds isolated from the methanol extract of the stem bark of *Antiaris africana* (Moraceae), for their antioxidant and antitumor activities using DPPH radical scavenging and XTT assays respectively [63]. Among these compounds was the phenolic compound 3,3'-dimethoxy-4'-*O*-β-*D*-xylopyronosylellagic acid (**41**), which showed the highest inhibition potency on both cell lines with more than 70% inhibition at 50 μg/mL thereby providing support for the traditional use of *A. africana* in the treatment of cancer.

Kuete *et al.* also isolated 2,2',5,6'-tetrahydroxybenzophenone (**43**) and isogarcinol (**44**) from the ethyl acetate and methanol extract of leaves of *Hypericum lanceolatum* (Clusiaceae) and isoxanthochymol (**45**) and guttiferone E (**46**) from the methanol extract of the stem bark of *Garcinia punctate* (Clusiaceae) [68]. The cytotoxicities of these compounds were evaluated on a resazurin reduction assay. The caspase-Glo assay was used to detect the activation of caspases 3/7, caspase 8 and caspase 9. Isogarcinol (**44**) was shown to be the most potent compound (IC_{50} ≤ 1 μg) against HCT116 colon carcinoma cells (p53+/+) (0.86 μM) and leukemia CCRF–CEM (1.38 μM) cell lines [68]. On the other hand, compounds **44 - 46** strongly induced apoptosis in CCRF–CEM cells. This lead to the validation of these plants for use in traditional medicine for the treatment of cancer/tumors [69]. The modes of action of compounds **43** to **46** have been elucidated. These compounds exhibit cytotoxicity against drug-sensitive and drug-resistant cancer cell lines. Compounds **44** to **46** were shown to induce apoptosis in CCRF–CEM cells through the activation of initiator caspases 8 and 9 and effector caspase 3/7 as well as loss of mitochondrial membrane potential. These compounds could therefore be explored in more detail in the future for the development of novel anticancer drugs against sensitive and resistant phenotypes [68].

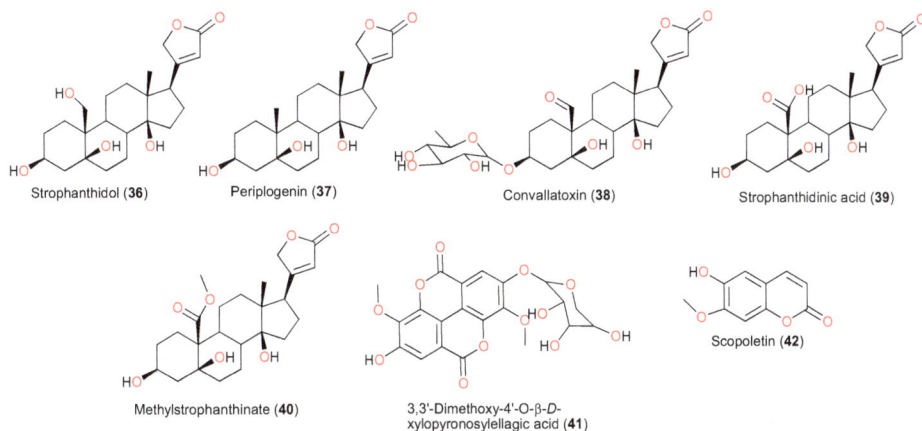

Fig. (6). Some cytotoxic compounds isolated from *A. africana* and *A. synopsis.*

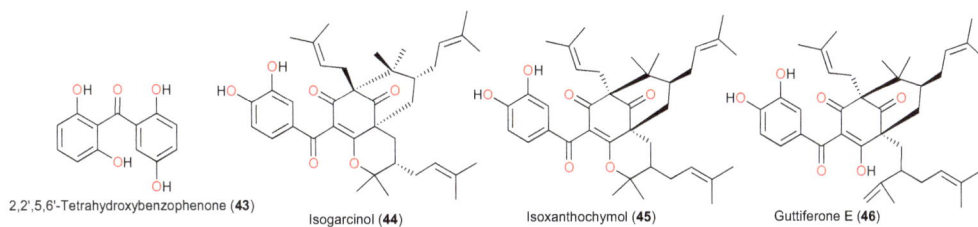

Fig. (7). Some cytotoxic compounds isolated from *H. lanceolatum* and *G. punctate.*

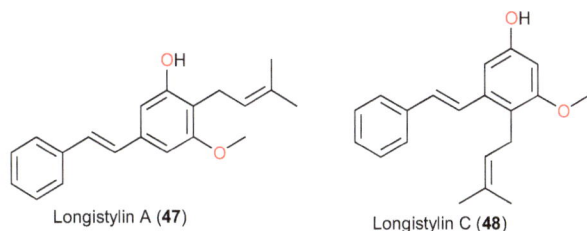

Fig. (8). Some cytotoxic compounds isolated from *C. cajan* (107, 108) and *T. africanus* (109, 110).

Ashidi *et al.* isolated two stilbenes, longistylins A (**47**) and C (**48**) from the MeOH extract of *Cajanus cajan* (Fabaceae) leaves as being primarily responsible for its cytotoxic activity for a range of cancer cell lines (IC$_{50}$ of 0.7 – 17.3 µM) [31]. The activities of these compounds were compared against the reference drug Vinblastine (IC$_{50}$ of 0.2 – 0.6 µM), supporting the use of the plant species (it is traditionally included in herbal preparations for treatment of cancer) [70].

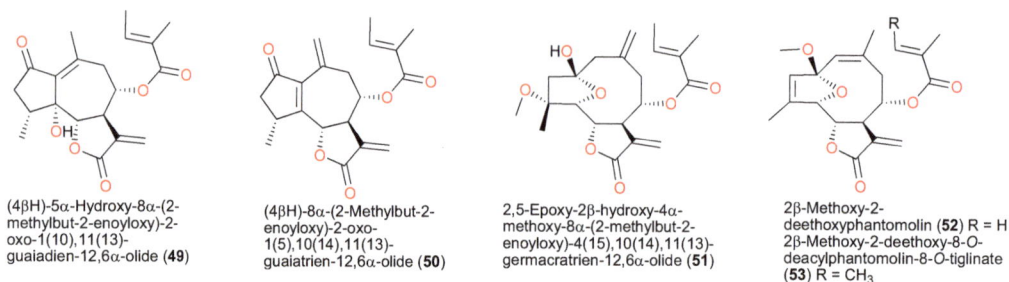

Fig. (9). Some cytotoxic compounds isolated from *Elephantopus mollis.*

Five sesquiterpene lactones (**49 – 53**) were isolated by Tabopda *et al.* from the roots and stems of *Elephantopus mollis* (Asteraceae). The isolated compounds were also evaluated for their cytotoxic activities against mouse neuroblastoma B104 cells [71]. All of the isolated compounds exhibited significant cytotoxic activities against the tested cell lines, with IC_{50} values which range from 1.58 to 3.85 μM.

CONCLUSION

This chapter summarizes the most important data for compounds isolated from African flora with a potential for anticancer drug discovery. For a more detailed information, the reader is referred to our most recent reviews [34, 35]. In general, to have certain cytotoxic activity in tumor is not sufficient to provide a compound with real antitumor activity and more important, a compound able to reach the market. Therefore, we would suggest that the compounds mentioned in this chapter be further investigated *in vivo*. Moreover, despite interesting initial results many of these compounds have not reach clinical testing. Arguably, there are many reasons for that, one of them being the absence of adequate screening facilities and human resources on the continent. In some cases, however, it is known that *in vitro* activity is often not translated to proper anticancer activity *in vivo* or in humans, or that the measured activity is not better than the drugs currently used for cancer treatments. In order to facilitate the process of lead/hit discovery of compounds with potential anti-cancer agents from the African flora by virtual screening, our team has recently developed a small natural products library, which has included about 400 plant metabolites from the African flora with promising anti-cancer properties [72]. This library has been analyzed for diversity and drug-likeness and compared with compounds currently included in the Naturally Occurring Plant-based Anti-cancer Compound-Activity-Target database (NPACT) library and with the Dictionary of Natural Products [73, 74]. The study also revealed that African medicinal plants constitute yet an untapped reservoir of lead compounds for anticancer drug discovery [72]. This is because the datasets from African flora (AfroCancer) and that from NPACT do not occupy the same chemical space [75] and there were discrepancies in the distribution of docking scores of the two datasets over a wide range of validated anticancer targets [72].

CONFLICT OF INTEREST

The authors confirm that they have no conflict of interest to declare for this publication.

ACKNOWLEDGEMENTS

CVS acknowledge a PhD scholarship from the German Academic Exchange Services (DAAD). FNK is currently a Georg Forster fellow of the Alexander von Humboldt Foundation, Germany.

REFERENCES

[1]   Torre LA, Bray F, Siegel RL, Ferlay J, Lortet-Tieulent J, Jemal A. Global cancer statistics, 2012. CA Cancer J Clin 2015; 65(2): 87-108.
        [http://dx.doi.org/10.3322/caac.21262] [PMID: 25651787]

[2]   American Cancer Society. Global Cancer Facts & Figures , 2011 [Accessed September 2014];1-60.
        http://d2j7fjepcxuj0a.cloudfront.net/wp-content/uploads/2014/02/2014_ACS_Statistics.pdf

[3]   Jemal A, Center MM, DeSantis C, Ward EM. Global patterns of cancer incidence and mortality rates and trends. Cancer Epidemiol Biomarkers Prev 2010; 19(8): 1893-907.
        [http://dx.doi.org/10.1158/1055-9965.EPI-10-0437] [PMID: 20647400]

[4]   Ferlay J, Shin HR, Bray F, Forman D, Mathers CD, Parkin D. GLOBOCA N 2008, Cancer incidence and mortality worldwide: IARC CancerBase No10. Lyon, France: International Agency for Research on Cancer 2010.

[5]   Kanavos P. The rising burden of cancer in the developing world. Ann Oncol 2006; 17 (Suppl. 8): i15-, i23.
        [http://dx.doi.org/10.1093/annonc/mdl983] [PMID: 16801335]

[6]   Rong G, Kang H, Wang Y, Hai T, Sun H. Candidate markers that associate with chemotherapy resistance in breast cancer through the study on Taxotere-induced damage to tumor microenvironment and gene expression profiling of carcinoma-associated fibroblasts (CAFs). PLoS One 2013; 8(8): e70960.
        [http://dx.doi.org/10.1371/journal.pone.0070960] [PMID: 23951052]

[7]   UK Cancer Research Worldwide Cancer Cancer Statistics (key Facts) Registered charity in England and Wales (1089464), Scotland (SC041666) and the Isle of Man (1103) Cancer Research UK 2014. Available at: http://publications.cancerresearchuk.org/downloads/product/cs_kf_allcancers.pdf

[8]   Chen EX, Siu LL. Development of molecular targeted anticancer agents: successes, failures and future directions. Curr Pharm Des 2005; 11(2): 265-72.
        [http://dx.doi.org/10.2174/1381612053382205] [PMID: 15638762]

[9] Early Breast Cancer Trialists Collaborative Group (EBCTCG). Effects of chemotherapy and hormonal therapy for early breast cancer on recurrence and 15-year survival: an overview of the randomised trials Lancet 2005; 365(9472): 1687-717.
[http://dx.doi.org/10.1016/S0140-6736(05)66544-0] [PMID: 15894097]

[10] Cragg GM, Newman DJ. Plants as a source of anti-cancer agents. J Ethnopharmacol 2005; 100(1-2): 72-9.
[http://dx.doi.org/10.1016/j.jep.2005.05.011] [PMID: 16009521]

[11] Goodman M, Morehouse F. Organic Molecules in Action. Gordon and Breach publishers 1973.

[12] Graham JG, Quinn ML, Fabricant DS, Farnsworth NR. Plants used against cancer - an extension of the work of Jonathan Hartwell. J Ethnopharmacol 2000; 73(3): 347-77.
[http://dx.doi.org/10.1016/S0378-8741(00)00341-X] [PMID: 11090989]

[13] Facts sheet 2004. Available at: http://www.who.int/mediacentre/news/releases/2004/pr44/en/

[14] Jansen O, Tits M, Angenot L, *et al.* Anti-plasmodial activity of *Dicoma tomentosa* (Asteraceae) and identification of urospermal A-15-*O*-acetate as the main active compound. Malar J 2012; 11: 289.
[http://dx.doi.org/10.1186/1475-2875-11-289] [PMID: 22909422]

[15] Merina N, Chandra KJ, Jibon K. Medicinal plants with potential anticancer activities : A review. Int Res J Pharm 2012; 3(6): 26-30.

[16] Sakarkar DM, Deshmukh VN. Ethnopharmacological review of traditional medicinal plants for anticancer activity. Int J Pharm Tech Res 2011; 3(1): 298-308.

[17] Lifongo LL, Simoben CV, Ntie-Kang F, Babiaka SB, Judson PN. A bioactivity *versus* ethnobotanical survey of medicinal plants from Nigeria, west Africa. Nat Prod Bioprospect 2014; 4(1): 1-19.
[http://dx.doi.org/10.1007/s13659-014-0005-7] [PMID: 24660132]

[18] Ntie-Kang F, Lifongo LL, Simoben CV, Babiaka SB, Sippl W, Mbaze LM. The uniqueness and therapeutic value of natural products from West African medicinal plants. Part I: Uniqueness and chemotaxonomy. RSC Advances 2014; 4: 28728-55.
[http://dx.doi.org/10.1039/c4ra03038a]

[19] Beutler JA, Cragg GM, Newman DJ. Drug Discovery in Africa impact of genomics, natural products, traditional medicines, insights into medicinal chemistry, and technology platforms in pursuit of new drugs. Berlin, Heidelberg: Springer Berlin Heidelberg 2012; pp. 29-52.

[20] Ntie-Kang F, Zofou D, Babiaka SB, *et al.* AfroDb: a select highly potent and diverse natural product library from African medicinal plants. PLoS One 2013; 8(10): e78085.
[http://dx.doi.org/10.1371/journal.pone.0078085] [PMID: 24205103]

[21] Gaulton A, Bellis LJ, Bento AP, *et al.* ChEMBL: a large-scale bioactivity database for drug discovery. Nucleic Acids Res 2012; 40(Database issue): D1100-7.
[http://dx.doi.org/10.1093/nar/gkr777] [PMID: 21948594]

[22] Batista R, Silva AdeJ Jr, de Oliveira AB. Plant-derived antimalarial agents: new leads and efficient phytomedicines. Part II. Non-alkaloidal natural products. Molecules 2009; 14(8): 3037-72.
[http://dx.doi.org/10.3390/molecules14083037] [PMID: 19701144]

[23] Dunkel M, Fullbeck M, Neumann S, Preissner R. SuperNatural: a searchable database of available natural compounds. Nucleic Acids Res 2006; 34(Database issue): D678-83.
[http://dx.doi.org/10.1093/nar/gkj132] [PMID: 16381957]

[24] Ntie-Kang F, Lifongo LL, Simoben CV, Babiaka SB, Sippl W, Mbaze LM. The uniqueness and therapeutic value of natural products from West African medicinal plants, part II: Terpenoids, geographical distribution, drug discovery. RSC Advances 2014; 4: 35348-70.
[http://dx.doi.org/10.1039/C4RA04543B]

[25]   Zofou D, Ntie-Kang F, Sippl W, Efange SM. Bioactive natural products derived from the Central African flora against neglected tropical diseases and HIV. Nat Prod Rep 2013; 30(8): 1098-120.
[http://dx.doi.org/10.1039/c3np70030e] [PMID: 23817666]

[26]   Idu M, Erhabor JO, Efijuemue HM. Documentation on medicinal plants sold in markets in Abeokuta, Nigeria. Trop J Pharm Res 2010; 9(2): 110-8.
[http://dx.doi.org/10.4314/tjpr.v9i2.53696]

[27]   Soh PN, Benoit-Vical F. Are West African plants a source of future antimalarial drugs? J Ethnopharmacol 2007; 114(2): 130-40.
[http://dx.doi.org/10.1016/j.jep.2007.08.012] [PMID: 17884314]

[28]   Kuete V, Sandjo LP, Ouete JL, Fouotsa H, Wiench B, Efferth T. Cytotoxicity and modes of action of three naturally occurring xanthones (8-hydroxycudraxanthone G, morusignin I and cudraxanthone I) against sensitive and multidrug-resistant cancer cell lines. Phytomedicine 2014; 21(3): 315-22.
[http://dx.doi.org/10.1016/j.phymed.2013.08.018] [PMID: 24075210]

[29]   Dzoyem JP, Nkuete AH, Kuete V, *et al.* Cytotoxicity and antimicrobial activity of the methanol extract and compounds from *Polygonum limbatum.* Planta Med 2012; 78(8): 787-92.
[http://dx.doi.org/10.1055/s-0031-1298431] [PMID: 22495442]

[30]   Sharma H, Parihar L, Parihar P. Review on cancer and anticancerous properties of some medicinal plants. J Med Plants Res 2011; 5(10): 1818-35.

[31]   Ashidi JS, Houghton PJ, Hylands PJ, Efferth T. Ethnobotanical survey and cytotoxicity testing of plants of South-western Nigeria used to treat cancer, with isolation of cytotoxic constituents from *Cajanus cajan* Millsp. leaves. J Ethnopharmacol 2010; 128(2): 501-12.
[http://dx.doi.org/10.1016/j.jep.2010.01.009] [PMID: 20064598]

[32]   Orang-Ojong BB, Munyangaju JE, Wei MS, *et al.* Impact of natural resources and research on cancer treatment and prevention: A perspective from Cameroon. Mol Clin Oncol 2013; 1(4): 610-20. [Review].
[PMID: 24649217]

[33]   Simoben CV, Ntie-Kang F, Lifongo LL, Babiaka SB, Sippl W, Mbaze LM. The uniqueness and therapeutic value of natural products from West African medicinal plants, part III: Least abundant compound classes. RSC Advances 2014; 4: 40095-110.
[http://dx.doi.org/10.1039/C4RA05376A]

[34]   Simoben CV, Ibezim A, Ntie-Kang F, Nwodo JN, Lifongo LL. Exploring cancer therapeutics with natural products from African medicinal plants, part I: Xanthones, quinones, steroids, coumarins, phenolics and others compound classes. Anticancer Agents Med Chem 2015; 15(9): 1092-111.
[http://dx.doi.org/10.2174/1871520615666150113110241] [PMID: 25584692]

[35]   Nwodo JN, Ibezim A, Simoben CV, Lifongo LL, Ntie-Kang F. Exploring cancer therapeutics with natural products from African medicinal plants, part II: Flavonoids, alkaloids and terpenoids. Anticancer Agents Med Chem 2016; 16(1): 108-27.
[http://dx.doi.org/10.2174/1871520615666150520143827] [PMID: 25991425]

[36]   Wansi JD, Devkota KP, Tshikalange E, Kuete V. Alkaloids from the medicinal plants of Africa. In: Kuete V, Ed. Medicinal plant research in Africa. Elsevier 2013.
[http://dx.doi.org/10.1016/B978-0-12-405927-6.00014-X]

[37]   Michael GS. Plant Systematics. London UK: WC1X 8RR 2006.

[38]   Cordell GA. Introduction to alkaloids: A biogenetic approach. New York: John Wiley & Sons 1981; pp. 450-62.

[39]   Barakat I, Jackson AH, Abdulla MI. Further studies of *Erythrina* alkaloids. Lloydia 1977; 40(5): 471-5.
[PMID: 73120]

[40]   Chacha M, Bojase-Moleta G, Majinda RR. Antimicrobial and radical scavenging flavonoids from the stem wood of *Erythrina latissima.* Phytochemistry 2005; 66(1): 99-104.
[http://dx.doi.org/10.1016/j.phytochem.2004.10.013] [PMID: 15649516]

[41]   Tanaka H, Sato M, Fujiwara S, Hirata M, Etoh H, Takeuchi H. Antibacterial activity of isoflavonoids isolated from Erythrina variegata against methicillin-resistant *Staphylococcus aureus*. Lett Appl Microbiol 2002; 35(6): 494-8.
[http://dx.doi.org/10.1046/j.1472-765X.2002.01222.x] [PMID: 12460431]

[42]   Mitscher LA, Drake S, Gollapudi SR, Okwute SK. A modern look at folkloric use of anti-infective agents. J Nat Prod 1987; 50(6): 1025-40.
[http://dx.doi.org/10.1021/np50054a003] [PMID: 3443855]

[43]   McKee TC, Bokesch HR, McCormick JL, *et al.* Isolation and characterization of new anti-HIV and cytotoxic leads from plants, marine, and microbial organisms. J Nat Prod 1997; 60(5): 431-8.
[http://dx.doi.org/10.1021/np970031g] [PMID: 9170286]

[44]   Njamen D, Mbafor JT, Fomum ZT, *et al.* Anti-inflammatory activities of two flavanones, sigmoidin A and sigmoidin B, from *Erythrina sigmoidea.* Planta Med 2004; 70(2): 104-7.
[http://dx.doi.org/10.1055/s-2004-815484] [PMID: 14994185]

[45]   Andayi AW, Yenesew A, Derese S, *et al.* Antiplasmodial flavonoids from *Erythrina sacleuxii.* Planta Med 2006; 72(2): 187-9.
[http://dx.doi.org/10.1055/s-2005-873200] [PMID: 16491458]

[46]   Mohammed MM, Ibrahim NA, Awad NE, *et al.* Anti-HIV-1 and cytotoxicity of the alkaloids of *Erythrina abyssinica* Lam. growing in Sudan. Nat Prod Res 2012; 26(17): 1565-75.
[http://dx.doi.org/10.1080/14786419.2011.573791] [PMID: 21936641]

[47]   Militão GC, Pinheiro SM, Dantas IN, *et al.* Bioassay-guided fractionation of pterocarpans from roots of *Harpalyce brasiliana* Benth. Bioorg Med Chem 2007; 15(21): 6687-91.
[http://dx.doi.org/10.1016/j.bmc.2007.08.011] [PMID: 17764956]

[48]   Cottiglia F, Casu L, Bonsignore L, *et al.* New cytotoxic prenylated isoflavonoids from *Bituminaria morisiana.* Planta Med 2005; 71(3): 254-60.
[http://dx.doi.org/10.1055/s-2005-837841] [PMID: 15770547]

[49]   Kuete V, Sandjo LP, Kwamou GM, Wiench B, Nkengfack AE, Efferth T. Activity of three cytotoxic isoflavonoids from *Erythrina excelsa* and *Erythrina senegalensis* (neobavaisoflavone, sigmoidin H and isoneorautenol) toward multi-factorial drug resistant cancer cells. Phytomedicine 2014; 21(5): 682-8.
[http://dx.doi.org/10.1016/j.phymed.2013.10.017] [PMID: 24252341]

[50]   Ndonsta BL, Tatsimo JS, Csupor D, *et al.* Alkylbenzoquinones with antiproliferative effect against human cancer cell lines from stem of *Ardisia kivuensis.* Phytochem Lett 2011; 4: 227-30.
[http://dx.doi.org/10.1016/j.phytol.2011.04.003]

[51]   Ndontsa BL, Tala MF, Talontsi FM, *et al.* New cytotoxic alkylbenzoquinone derivatives from leaves and stem of *Ardisia kivuensis* (Myrsinaceae). Phytochem Lett 2012; 5: 463-6.
[http://dx.doi.org/10.1016/j.phytol.2012.04.006]

[52]   Paul DJ, Laure NB, Guru SK, *et al.* Antiproliferative and antimicrobial activities of alkylbenzoquinone derivatives from *Ardisia kivuensis.* Pharm Biol 2014; 52(3): 392-7.
[http://dx.doi.org/10.3109/13880209.2013.837076] [PMID: 24192208]

[53]   Krief S, Thoison O, Sévenet T, Wrangham RW, Lavaud C. Triterpenoid saponin anthranilates from *Albizia grandibracteata* leaves ingested by primates in Uganda. J Nat Prod 2005; 68(6): 897-903.
[http://dx.doi.org/10.1021/np049576i] [PMID: 15974615]

[54]   Fatope MO, Audu OT, Takeda Y, *et al.* Bioactive *ent*-kaurene diterpenoids from *Annona senegalensis.* J Nat Prod 1996; 59(3): 301-3.
[http://dx.doi.org/10.1021/np9601566] [PMID: 8882434]

[55]   Chiang YM, Kuo YH, Oota S, Fukuyama Y. Xanthones and benzophenones from the stems of *Garcinia multiflora.* J Nat Prod 2003; 66(8): 1070-3.
[http://dx.doi.org/10.1021/np030065q] [PMID: 12932126]

[56]   Ito C, Itoigawa M, Takakura T, *et al.* Chemical constituents of *Garcinia fusca*: structure elucidation of eight new xanthones and their cancer chemopreventive activity. J Nat Prod 2003; 66(2): 200-5.
[http://dx.doi.org/10.1021/np020290s] [PMID: 12608849]

[57]   Matsumoto K, Akao Y, Kobayashi E, *et al.* Induction of apoptosis by xanthones from mangosteen in human leukemia cell lines. J Nat Prod 2003; 66(8): 1124-7.
[http://dx.doi.org/10.1021/np020546u] [PMID: 12932141]

[58]   Mackeen MM, Ali AM, Lajis NH, *et al.* Antimicrobial, antioxidant, antitumour-promoting and cytotoxic activities of different plant part extracts of *Garcinia atroviridis* griff. ex T. anders. J Ethnopharmacol 2000; 72(3): 395-402.
[http://dx.doi.org/10.1016/S0378-8741(00)00245-2] [PMID: 10996278]

[59]   Thoison O, Fahy J, Dumontet V, *et al.* Cytotoxic prenylxanthones from *Garcinia bracteata.* J Nat Prod 2000; 63(4): 441-6.
[http://dx.doi.org/10.1021/np9903088] [PMID: 10785410]

[60]   Azebaze AG, Menasria F, Noumi LG, *et al.* Xanthones from the seeds of *Allanblackia monticola* and their apoptotic and antiproliferative activities. Planta Med 2009; 75(3): 243-8.
[http://dx.doi.org/10.1055/s-0028-1088375] [PMID: 19053018]

[61]   Raponda–Walker A, Sillans R. Les plantes utiles du Gabon. Paris: Paul Lechevalier 1961; Vol. 6.

[62]   Tala MF, Tchakam PD, Wabo HK, *et al.* Chemical constituents, antimicrobial and cytotoxic activities of *Hypericum riparium* (Guttiferae). Rec Nat Prod 2013; 7(1): 65-8.

[63]   Kuete V, Vouffo B, Mbaveng AT, Vouffo EY, Siagat RM, Dongo E. Evaluation of *Antiaris africana* methanol extract and compounds for antioxidant and antitumor activities. Pharm Biol 2009; 47: 1042-9.
[http://dx.doi.org/10.3109/13880200902988595]

[64]   Berhaut J. Flore Illustrée du Sénégal. Paris: Satabié 1979; pp. 402-5.

[65]   Okogun IJ, Spiff AI, Ekong DE. Triterpenoids and betaines from the latex and bark of *Antiaris africana.* Phytochemistry 1976; 15: 826-7.
[http://dx.doi.org/10.1016/S0031-9422(00)94464-9]

[66]   Happi EN, Tcho AT, Sirri JC, *et al.* Tirucallane triterpenoids from the stem bark of *Araliopsis synopsis.* Phytochem Lett 2012; 5: 423-6.
[http://dx.doi.org/10.1016/j.phytol.2012.03.014]

[67]   Irvine FR. Woody plants of Ghana. Oxford: Oxford University Press 1961; p. 868.

[68]   Kuete V, Tchakam PD, Wiench B, *et al.* Cytotoxicity and modes of action of four naturally occuring benzophenones: 2,2,5,6-tetrahydroxybenzophenone, guttiferone E, isogarcinol and isoxanthochymol. Phytomedicine 2013; 20(6): 528-36.
[http://dx.doi.org/10.1016/j.phymed.2013.02.003] [PMID: 23507522]

[69]   Focho DA, Ndam WT, Fonge BA. Medicinal plants of Aguambu - Bamumbu in the Lebialem

highlands, South West province of Cameroon. Afr J Pharm Pharmacol, 2009; 3: 001-13.

[70] Akinsulie AO, Temiye EO, Akanmu AS, Lesi FE, Whyte CO. Clinical evaluation of extract of *Cajanus cajan* (Ciklavit) in sickle cell anaemia. J Trop Pediatr 2005; 51(4): 200-5.
[http://dx.doi.org/10.1093/tropej/fmh097] [PMID: 15917266]

[71] Tabopda TK, Ngoupayo J, Liu J, *et al.* Further cytotoxic sesquiterpene lactones from *Elephantopus mollis* KUNTH. Chem Pharm Bull (Tokyo) 2008; 56(2): 231-3.
[http://dx.doi.org/10.1248/cpb.56.231] [PMID: 18239317]

[72] Ntie-Kang F, Nwodo JN, Ibezim A, *et al.* Molecular modeling of potential anticancer agents from African medicinal plants. J Chem Inf Model 2014; 54(9): 2433-50.
[http://dx.doi.org/10.1021/ci5003697] [PMID: 25116740]

[73] Mangal M, Sagar P, Singh H, Raghava GP, Agarwal SM. NPACT: Naturally Occurring Plant-based Anti-cancer Compound-Activity-Target database. Nucleic Acids Res 2013; 41(Database issue): D1124-9.
[http://dx.doi.org/10.1093/nar/gks1047] [PMID: 23203877]

[74] Ntie-Kang F, Lifongo LL, Judson PN, Sippl W, Efange SM. How drug-like are naturally occurring anti-cancer compounds? J Mol Model 2014; 20(1): 2069.
[http://dx.doi.org/10.1007/s00894-014-2069-z] [PMID: 24452907]

[75] Ntie-Kang F, Simoben CV, Karaman B, *et al.* Pharmacophore modeling and *in silico* toxicity assessment of potential anticancer agents from African medicinal plants. Drug Des Dev Ther 2016; 10: 2137-54.
[http://dx.doi.org/10.2147/DDDT.S108118] [PMID: 27445461]

CHAPTER 6

Phytochemicals in Therapy of Radiation Induced Damage and Cancer

Pankaj Taneja[1,*], Mehak Gulzar[1], Neetu Kumra Taneja[2], BS Dwarakanath[3] and RP Tripathi[4]

[1] *Department of Biotechnology, School of Engineering and Technology, Sharda University, Knowledge Park-III, Gautam Buddha Nagar, Greater Noida, Uttar Pradesh, India, Consultant, Nutrametrix Health Solutions, USA*

[2] *National Institute of Food Technology Entrepreneurship and Management, Kundli, Sonipat, Haryana, India*

[3] *Central Research Facility, Sri Ramachandra University, Porur, Chennai 600116, India*

[4] *Institute of Nuclear Medicine and Allied Science, SK Majumdar Marg, New Delhi, India*

Abstract: Radiation has been implicated in causing deleterious effect including cancer. Radiation exposure cause mutations, damage to the hematopoietic, gastrointestinal or central nervous systems which are critical causing adverse health effects. Hence, there is an urgent need to prevent such effects. A majority of phytochemicals have potential chemopreventive efficacy with relative less toxicity. Specifically, the utilization of these natural plants as modifiers of the radiation reaction is accepting extensive considerable attention. In this current review, we summarize the antimutagenic and anticancer effects of some selected natural phytochemicals including Amaranthus, Bael, Angelica, Rhamnoides, Haloil, Ginseng, Moringa, Biophytum *etc* against radiation induced damage.

Keywords: Cancer, Phytochemicals, Radiation damage, Therapy.

INTRODUCTION

Radiation harmful reactions of cells are described by damage to biomolecules including DNA lesions, perturbations in cell cycle, cytogenotoxicity, dysregulation of cell signaling and gene expression [1 - 3]. Depending on dose levels and duration of radiation exposure, these can cause mutation, damage to the hematopoietic, gastrointestinal or central nervous systems and cancer [1 - 3].

---

* **Corresponding author Pankaj Taneja:** Department of Biotechnology, School of Engineering and Technology, Sharda University, Knowledge Park-III, Gautam Buddha Nagar, Greater Noida, Uttar Pradesh, India; Tel: +91-9560813083; USA-3366085675; E-mails: pankajtnj@gmail.com, pankaj.taneja@sharda.ac.in

Sahdeo Prasad & Amit Kumar Tyagi (Eds.)

We also see that radiation is accounted for therapeutic mortality for the treatment of solid tumors,but its exposure to surrounding normal cells leads to mutation which results in the occurrence of cancer [3, 4]. Moreover, events like the Hiroshoma Nagasaki atomic bomb explosion showed incidences of cancer due to those radiation exposed to populations and to its generations. Hence the radiation protection countermeasures is an urgent need. Using natural products as radioprotectors is of great interest because of their relatively low toxicity and ability to develop better drugs [5, 6]. Focus on plant research has increased in recent times with an aim towards their edible, medicinal and amelioration properties in animal welfare globally. Bunches of proof have gathered to indicate gigantic capability of medicinal nutritional plants in ethical society. In the course of the most recent years, researcher scientists have gone for recognizing, identifying, approving and validating plant inferred substances for disease treatment and combating toxicity. It has been demonstrated that different parts of plants, for example, leaves, organic products, seeds give wellbeing and nourishment in the human diet which has enormous traditional use against various diseases [4 - 6]. These phytochemicals indicate efficient anticancer activity. They have cancer prevention properties with typical features of protecting normal cells from radiation damage. They also exhibit prooxidant function in cancer cells, which increases the damage imposed by radiation [5, 6]. There exists a balance between pro- or antioxidant function of these phytochemicals which is concentration dependent and regulated by cytosolic redox status. The use of these phytochemicals as radio-sensitizing agents is picking up force suggesting it be chemomodulator of radiotherapy making them effective curative agents.We summarize in this review few studies on phytochemicals discussed below for their use as radioprotectors. Most of these herbs are of Indian and Chinese origin because they are one of the most ancient countries using traditional natural products (Text below and Table **1**).

Table 1. Radioprotective effects of few Phytochemicals in cancer.

Phytochemical	Protection	Cancer	Origin
Aegle marmelos Corr.ex.Roxb.	protection against radiation induced sickness, human peripheral blood lymphocytes (HPBLs) irradiated with gamma-radiation	breast cancer	Rutaceae, cultivated in north India
Acanthopanax senticosus	radioprotective effects on hemopoiesis of irradiated mice, Pre-irradiation administration of Shigoka extract (5 mg/kg b.w.; −24 h; i.p.) rendered maximum survival (80%), while post-irradiation administration (+12 h; 9.5 Gy) exhibited 30% survival.	cerebral haemorrhage	Araliaceae

(Table 1) contd.....

Phytochemical	Protection	Cancer	Origin
Amaranthus paniculatus Linn.	Oral administration significant decrease in tumor volume, viable cell count and tumor weight	Ehrlich's ascites carcinoma	Amaranthaceae
Angelica sinensis	effective down-regulation of TNF-· and TGF-ß1 irradiated lung tissue	Lung cancer	Apiacae
Aloe Vera & Aloe barbadenis	contains emodin that activates the macrophages to fight cancer, inhibit metastases	Merkel cell carcinoma	Liliaceae
Biophytum sensitivum	reduce the enhanced level of ALP, GPT and LPO levels and significantly enhance the glutathione (GSH) content in liver and intestinal mucosain gamma irradiated animals,	Liver cancer	Oxidaceae
Centella asiatica	Aqueous extracts protects against low dose ionization radiation, administrative orally total body against sublethal gamma radiation	Whole body	Apiaceae
Coronopus didymus	Optimum radioprotection was observed upon *i.p.* administration, 30 min prior to 10 Gy irradiation	Whole body	Brassicaceae
Emblica officinalis	effectively prevent gamma ray–induced lipid peroxidation	*In vitro*	Euphorbiaceae
Glycyrrhiza glabra	Protection against microsomal membrane gamma radiation induced lipid peroxidation	*In vivo* (rats)	Leguminosae
Hypercium perforatum	Management of acute skin toxicity in head and neck cancer patients undergoing radiation	Head and neck cancer	Hyperiaceae
Hippophae rhamnoides	Dose of 30 mg/kg body weight of RH-3 rendered 82% survival	*In vitro*	Elaegnaceae
Mentha arvensis	Extract provides protection against the radiation-induced sickness and mortality and the optimum protective dose of 10 mg/kg is safe from the point of drug-induced toxicity.	Mice	Laminacaeae
Moringa oleifera	Hepatoprotective effect, pretreatment with MoLE protected against γ-radiation-induced liver damage.	Liver cancer	Moringaceae
Panax ginseng	Protection against gamma radiation, Lipid peroxidation, glutathione,chromosomal damage	*In vivo / in vitro*	Aralaceae
Tephrosia purpurea	Tephrosia extract (200 mg/kg b.wt) protected Swiss albino mice against radiation (5 Gy)-induced haemopoietic injury	Mice	Fabaceae

Amaranthus Paniculatus

Amaranthus paniculatus(Amaranthaceae) has Indian name as "Rajgira" and "Amaranth" as English name. It is cultivated as traditional medicinal plant by native Americans and other parts of world. The plant has widespread use in folk and traditional medicines for respiratory infections, vision defects, tuberculosis, fleshy tumors, liver complaints and inflammations. Decocting leaves of plant are used for treating chest pain, gastroenteritis and seeds are applied to sores in ayurvedic preparations. Radioprotection has been observed by extract from leaves of this plant against whole body gamma radiation [7]. It is likewise overcome the problems of psychological stress; its impact has been tried in protecting radiation stress induced memory dysfunction [8]. Amaranthus paniculatus leaf extracts have content of lysine, methionine, carotenoids, Vitamin C and proteins [9]. Amaranthus paniculatus has been found to exhibit antioxidant properties against Ehrlich's ascites carcinoma (EAC) in mice [10]. It regulates cellular antioxidant defense system such as catalase, superoxide dismutase and glutathione, suggesting chemopreventive effect of Amaranthus against oxidative stress conditions which induce detoxification mechanisms.

Bael (Aegle Marmelos)

The Bael (Aegle marmelos) (L) Corr. is an ancient Indian medicinal drug used by sacred groves for the treatment of various diseases due to its medicinal properties including antigenotoxic and anticancer effects [11]. The radio-protective effect of flavonoids was due to the scavenging of radiation induced free radicals [12]. Bael leaf extract when given at the dose of 15mg/kgbw to mice increased glutathione levels and reduced lipid peroxidation caused by radiation [11, 12]. Irradiation causes sickness and mortality. It results elevated levels of hepatic, renal, stomach and intestine lipid peroxidation and well as depleted in GSH levels. [11, 12]. Treatment of Aegle leaf extract at concentration of 5µg/ml in cultured human peripheral blood lymphocytes (HPBLs) irradiated with different doses of gamma-radiation leads to decreased radiation induced micronuclei formation [13]. It is also found to inhibit *in vitro* proliferation of human tumour cell lines K562, T-lymphoid jurhat, Beta lymphoid Raji, & HEL [14]. Bael leaves shows the antitumor effect in the animal model of Ehrlich ascites carcinoma in mice at dose of 400 mg/kg bw [15]. The plant extract also exhibits cytotoxicity against tumor cell lines in brine shrimp lethality and methyl thiazolyl tetrazolium (MTT) based assays [15]. Bael has been found to possesses anti-proliferative activity on MCF7 and MDA-MB-231 breast cancer cell lines [16].

Angelica Sinensis

Angelica Sinensis, ('Danggui') is a Chinese medicine, used widely treat radiation-induced pneumonitis. Radiation therapy in lung cancer patients has side effects of the such as pneumonitis and excessive fibrosis [17]. Hua investigated that

Angelica Sinensis treatment with and without X-ray irradiation (12Gy) resulted in modulation of immunological cytokines [18]. Angelica Sinensis(AS) administration in response radiation exposure to of X-ray irradiation to thoraces at 12Gy exhibited low lower mRNA levels levels of TNF-α and TGF-ß1 in no treatment [NT] AS treated group in mice [18]. In XRT mice, there were increased inflammatory cells positive for TNF-α and TGF-ß1 in the lung tissue compared with NT mice. The persistent localization and infiltration of TNF-α and TGF-ß1 suggests the involvement of cytokines' in the process of radiation-induced damage. Besides, powerful down-control of TNF-α and TGF-ß1 in irradiated lung tissue by Angelica Sinensis is demonstrative of its clinical efficacy in treating radiation-actuated damage. Angelica extract has been utilized in treatment of cancer patients with radiation-induced pneumonitis as an empirical clinical practice of Chinese medicine, with low/no toxicity [19].

H. Rhamnoides

H. rhamnoides L. (*F. Elaeagnaceae*), commonly known as sea buckthorn and native to Europe & Asia has been found to exhibit anti-oxidative, immunostimulative activity [20]. Its radioprotective potential has also been investigated [21]. *H. rhamnoides* contains vitamin A, C, E and K, tannins and flavonoids, certain trace elements like Se, Zn, Cu and S which are the part of metallo-enzymes and some of which exhibit its antioxidant and radioprotection potential [22]. Goel *et al.* [23] administered rhamnoides extract through i.p. route at 30mg/kgbw to the mice 30 minutes before whole body irradiation. The animals were studied for analysing body survival, spleen colony forming units (CFU) and hematological parameters. RH-3 produced 82% survival in comparison to no survival in irradiated control. Endogenous splenic CFU counts on 10th post-irradiation day with and without RH-3 inhibited radiation mediated generation of hydroxyl radicals *in vitro*. It was accompanied by free radical scavenging, acceleration of stem cell proliferation and immuno-stimulation demonstrating radioprotective attributes.

Holoil

Holoil is derived from hypericum flowers (*Hypericum perforaturm*) and neem oil (*Azadirachta indica*) extracts. Hypericum flowers and Neem oil have been demonstrated to have anti-inflammatory properties [24 - 26]. Hyperforin component of hypericum reduces *in-vivo* neo-vascularization; inhibit angio-genesis, invasion, endothelial cell proliferation and differentiation in murine model. It also modulates MMP-2 and urokinase-mediated extracellular matrix degradation [27]. Neem oil is a vegetable oil got by icy extraction from the berries of Azadirachta indica, an evergreen tree endemic to the Indian subcontinent. Neem extricates have been utilized for a considerable length of time as cosmetics and cicatrizing, bacteriostatic and anti-inflammatory agents in Indian traditional medicine [27]. It is a significant part of the Ayurvedic Pharmacopoeia India [26, 28]. The inflammatory protection properties of neem oil are because of the presence of a limonoid (epoxyazadiradione) which acts by inhibition of macrophage migration [29]. Holoil is used in the treatment of intense skin toxicity in patients undergoing radiotherapy or chemo-radiotherapy for head and neck tumors by Francol *et al.*, [30] demonstrated protective potential of Holoil in twenty eight head and neck cancer patients getting radiotherapy (RT) in mono-institutional single-arm prospective observational study. Patients receiving both definitive or post-operative radiotherapy were permitted, with options of exclusive modality or combined with (concomitant or induction) chemotherapy [30]. Holoil was administered to patients with bright erythema, moderate oedema or patchy moist desquamation. Holoil was used during all RT procedure with follow up time, until acute skin toxicity recovery was achieved. A prophylactic effect in the prevention of moist desquamation was observed with Holoil, proving to be a safe and active option in the management of acute skin toxicity in head and neck cancer patients receiving RT or chemo-radiotherapy.

Ginseng

Ginseng roots belongs to the species of the genus Panax. It has been widely used as a medicinal herb in traditional Chinese medicine as tonic, immunomodulatory, antimutagenic, adaptogenic and antiaging agent [31, 32]. Ginseng medicinal effects are attributed to the triterpene glycosides known as ginsenosides (saponins) [33 , 34]. Ginseng is attributed to exhibit antineoplastic and other pharmacological beneficial activities [35 - 38]. In addition, ginseng and its partially purified constituents have potential radioprotective properties [39]. Ginseng saponin purified mixture provides resistance against γ-irradiation in the *ex vivo* colony forming assay in V-79 cells [40]. P.ginseng water extract pre treatment for 48hrs reduces frequency of DNA double strand breaks in cultured spleen lymphocytes induced by gamma radiation [41]. It reduces t radiation-induced apoptosis in both jejunal crypt cells and hair follicles [42]. Chinese herbal medicine containing 25% of *P.ginseng* root extract enhances radiotolerance (X-ray exposure from 0–5 Gy) of bone marrow stem cells and peripheral hematocytes [43]. Moreover, in Swiss albino mice, *P.ginseng* extract significantly decreases testicular acid phosphatase activity and lipid peroxidation levels demonstrating protection against γ ray-induced testicular damage [44]. Aqueous extract of whole ginseng provides a better protection against radiation-induced DNA damage as compared to isolated ginsenoside fractions. Since free radicals causes radiation-induced damage, the underlying radioprotective mechanism of ginseng could be linked, either directly or indirectly, to its antioxidative capability by the scavenging free radicals responsible for DNA damage. In addition, ginseng's radioprotective potential may also be attributed through its immunomodulating capabilities. Ginseng is a natural product with worldwide distribution, and in addition to its antitumor properties, ginseng is also promising radioprotector for therapeutic or preventive protocols capable of attenuating the deleterious effects of radiation on human normal tissue, especially for cancer patients undergoing radiotherapy.

Moringa Oleifera

Moringa oleifera Lam. (synonym Moringapterygosperma Gaertn., family Moringaceae) tree has several beneficial medicinal effects. *M. oleifera* leaves are good source of natural antioxidants. such as ascorbic acid, flavonoids, phenolics, and carotenoids.The high concentrations of ascorbic acid, b-sitosterol, iron, calcium, phosphorus, copper, vitamins A, B, and C, tocopherol, riboflavin, nicotinic acid, folic acid, pyridoxine, b-carotene, protein, and essential amino acids (methionine, cysteine, tryptophan, and lysine) present in Moringa leaves make a valuable dietary supplement [45]. It also has hepatoprotective activity [46, 47]. Moringa leaf extract prevents bone marrow chromosomes from radiation-induced damage [48]. Aqueous ethanolic Moringa oleifera leaf extract (MoLE) has also been shown to reduce radiation induced oxidative stress inflammation and lipid peroxidation [49]. The protection was observed due to the free radical scavenging activity of MoLE, which can ameliorate radiation-induced oxidative stress.

Biophytum Sensitivum

Biophytum sensitivum(L.) DC. (Family-Oxalidaceae) is a medicinal plant used in traditional medicine by population in Asia, Africa, and the Pacific islands. [50, 51]. It has been documented to exhibit anti-inflammatory and antidiabetic effects [52, 53]. *Biophytum sensitivum* also possess antioxidant, anti-angiogenic and immunomodulatory and antitumor potential [54 - 56]. Radioprotective effect of methanol extract of Biophytum sensitivum has been shown in mice exposed to whole body gamma irradiation (6 Gy/ animal) after treatment with *B. sensitivum* (50mg/kg b.wt.) [57]. It was found to reduce the enhanced level of ALP, GPT and LPO levels in irradiated animals. *B. Sensitivum* could significantly enhance the glutathione (GSH) content in liver and intestinal mucosa of irradiated animals. Its treatment could enhance the total WBC count, cellularity of bone marrow, alpha-esterase positive cells, and relative organ weight of spleen as well as thymus [57]. *B. sensitivum* treatment can activate cytokines such as IL-1β, IFN-γ and GM-CSF in animals exposed to whole body gamma irradiation. Immunomodulation and induction of IL-1β, GM-CSF and IFN-γ by *Biophytum sensitivum* against radiation induced hemopoietic damage are the mechanisms of its protection.

Acanthopanax Senticosus

Acanthopanax senticosus is also named Eleutherococcus senticosus, and Ciwujia is a Chinese herbal product. It has medicinal activity in comprising radio-protective, anti-inflammatory and hepatoprotective pharmacological properties [58]. Miyanomae and Frindel [59] conducted study on Acanthopanax Shigokal extract. They showed radioprotective effects of this extract on hemopoiesis of irradiated mice, it was incurred by the recovery of spleen colony-forming units at dose upto the level of 7Gy [59]. Minor anticarcinogenic activity of Eleutherococcus is also reported in γ-irradiation induced breast tumors in rats [60].

Aloe Vera; Barbadensis

Aloe belongs to family Alliaceae, which is a succulent perennial pea green herb having fleshy serrated leaves origin in Southern and Eastern Africa. Several species *Aloe vera* (L.), Burm. *f.* syn. *Aloe barbadensis* is the most biologically active among 400 different species. It is also one of the traditional herbs good for skin and also possessing antioxidant and anticancer activity [61, 62]. Radioprotective efficacy of Aloe as scavengers of free radicals and DNA damage inhibition is well documented [63]. Extracts oligosaccharides of crude Aloe barbadensis in gel form suppress UV light induced T cell mediated delayed type hypersensitivity (DTH) by regulating reduction in the release of imm-unoregulatory cytokine IL-10 in the skin of mice [64]. In another study, animals pre-treated with Aloe juice 0.25ml/kgbw/day for 5days, were exposed to 7Gy γ radiation, showed better antioxidant status in various tissues [65]. Aloe juice inhibits radiation-induced increment of malondialdehyde in the liver, lungs, and kidney tissues of irradiated rats. Significant amelioration in superoxide dismutase (SOD) and catalase activities was also identified [65]. One of the carbohydrates (Acemannan) is considered as the main functional component of *Aloe vera*, which is known to promote wound healing and protect radiation induced skin damage [62, 66]. Aloe gel and creams are clinically proven to heal the radiation burns in patients [67 - 69]. *Aloe vera* juice promotes body to heal itself from cancer and also from the genotoxic damage caused by radio and chemotherapy by boosting the immune system [62, 67 - 69]. *Aloe vera* emodin, an anthraquinone, has the ability to suppress the growth of malignant cancer cells thus exhibiting anti-tumor properties as well [70, 71].

Mentha

Mentha (also known as mint) is a genus of plants in the family Lamiaceae [72]. Mentha plant extract has also the potential of radioprotector efficacy [73]. The choloroform extract of Mentha at doses from 10mg/kg to 80mg/kgbw for 30 days was found to exhibit protection against radiation damage caused by 7Gy dose in mice [74]. Mentha improved the health of animals against the gastrointestinal death as well as bone marrow deaths. Moreover, Mentha alone was found to be non-toxic up to a dose of 1,000 mg/kgbw [74]. M. piperita when given at the dose of (1g/kgbw/day) for 20days showed intestinal mucosa protection after whole body 8 Gy gamma irradiation exposure [75]. Hassan *et al.* [76] demonstrated that Mentha can ameliorate the neuronal injury induced by gamma irradiation. They also found that Mentha down-regulated P53 expression and up-regulation of Bcl(2) domain protected brain structure from extensive damage in response radiation. Also it has been found that Mentha extract possesses antioxidant potential against radiation induced free radicals. In response to 8.0 Gy gamma radiation exposure, Mentha treatment showed a significant increase in the activities of glutathione peroxidase, glutathione-S-transferase(GST), superoxide dismutase and catalase [77].

Coronopus Didymus

Coronopus didymus belongs to the species of Coronopus, which is a genus of plants in the Brassicaceae (mustard family) with medicinal properties in various ailments [78]. Chrysoeriol and its glycoside (chrysoeriol-6-O-acetyl-4'-beta-D-glucoside) are extracted from the *Coronopus didymus* [78]. They reduce superoxide anion by xanthine/xanthine oxidase system and also exhibit antioxidant effect against radiation [78]. Radioprotection of its aqueous extract was found by injection in the mice before exposure to gamma radiation at a range from 7-11Gy [79]. Endogenous lipid peroxidation damage by radiation and antioxidant levels were recovered in free radical scavenging fraction (CDF1), fraction of Coronopus treated group [79].

Centella Asiatica

Centella asiatica commonly known as *mandukaparni* or Indian pennywort or *jalbrahmi* is used in Ayurvedic traditional medicine for thousands of years, it is enlisted in Indian and Indonesian medical text as 'Sushruta Samhita' and as gotukola in China [80 , 81]. *Centella asiatica* (aqueous extract) was tested for its efficacy against radiation induced damage to behavioral perturbations like conditioned taste aversion (CTA), performance decrement and learning, Administration of *C. asiatica* at the dose of 100mg/kg.bw rendered significant inhibition against radiation-induced body weight loss and CTA that became evident on the second post irradiation [82].

This plant extract protects against cellular membrane damage caused by radiation [83]. The titrated extract of *Centella asiatica* (TECA) is a modified mixture containing of asiatic acid, madecassic acid, asiaticoside and madecassoside. It exhibits therapeutic potential in wound healing acts as an anti-microbial, anticancer and anti-aging agent [80 , 84]. The TECA extract was found to exhibit protection against UVB induced alterations in miRNA expression [84].

Glycyrrhiza Glabra

Glycyrrhiza glabra(liquorice) belongs to the pea and bean family Leguminosae/ Fabaceae is sweet Greek root. It is used in making liquorice-flavoured confectionery. It is cultivated for its rhizomes (underground stems) that contain the compound glycyrrhizin, which is used in medicine as an alternative sweetener. Medicinally, it is used as an ingredient in cough mixtures and throat lozenges, to treat sore throats, mouth ulcers, stomach ulcers, inflammatory stomach conditions and indigestion [85]. Glabridin, present in hydrophobic fraction, inhibits UVB induced pigmentation, erythema, inflammation and tyrosinase activity in skin [85]. Glycyrrhiza glabra can be used effectively for the prevention and treatment of oral mucositis post radiation and chemotherapy in head and neck cancer patients [86]. Licochalcone A (LicoA) another compound in G. glabra root has been shown to suppress ultraviolet induced COX2 expression, prostaglandin E2 expression and AP-1 transcriptional activity [87]. LicoA reduced UV -induced phosphorylation of Akt/ mammalian target of rapamycin (mTOR) and extracellular signal-regulated kinases (ERK)1/2/p90 ribosomal protein S6 kinase (RSK) in HaCaT cells [87].

Tephrosia Purpurea

Tephrosia purpurea is a flowering plant of the pea family (Fabaceae). It has a pantropical distribution as wasteland weed. It is also cultivated as green manure crop in various parts of world including India and Sri Lanka. It exhibits anthelmintic, alexiteric, restorative, and antipyretic properties in Ayurveda. It is used for treating leprosy, ulcers, asthma and cancer and various disorders in liver, spleen, heart, and blood [88 - 91]. Tephrosia extract inhibits radiation induced haemopoietic injury when given at the dose of 200mg/kgbw and is a potent scavenger of free radicals with antioxidant activity [90 - 93].

Emblica Officinalis

Emblica officinalis belongs to family Phyllanthaceae called asemblic myrobalan, myrobalan, Indian gooseberry, Malacca tree or amla [94]. It has also been reported to possess nutritional and pharmacological therapeutic properties since ancient times [94]. Emblica fruit extract protects against ultraviolet-B (UVB) induced reactive oxygen species (ROS) and collagen damage in normal human dermal fibroblasts [95]. It is also used as a natural active cosmetic agent against photo-aging [96]. Administration of Emblica officinalis extract (EOE) in mice exposed to 5Gy gamma radiation showed protection by depletion in lipid peroxidation and elevation in glutathione and catalase levels [97]. Pre-treatment of Embolic officinal is protects against gamma radiation induced hematological and biochemical damage in peripheral blood [98]. Significant restoration of RBC, WBC, hemoglobin, and hematocrit was observed in E. officinal is group as compared to irradiated group [98]. When given at the dose of 100mg/kgbw, Emblica has shown to reduce upto 9Gy radiation induced damage and mortality [99]. The antitumorigenic potential of Emblica has been observed in skin and ovarian cancer [100, 101].

Clinical Trials

Some phytochemicals have also been found to have substantial radiosensatisation and anticancer activity in human clinical trials. *Eleutherococcus* (Acanthopanax) exhibited immuno-stimulating properties in radiation in breast cancer patients. *In vitro* treatment of lymphocytes with eleutherococcal preparation produced an immune-boosting effect as the mechanism of therapy in these cancer patients [102]. Angelica sinensis has also been found to increase the immune cell count of cancer patients receiving chemotherapy and/or radiotherapy to prevent leucopenia [103]. Ginseng radioprotective potential was observed in one of the clinical studies [104]. North American ginseng extract (NAGE, total ginsenoside content: 11.7%) was given at the dose of 250-1000 ug/ml at 90 minutes postirradiation of 1 and 2 Gy to 40 individuals. DNA damage and oxidative stress were measured in their lymphocytes. NAGE (750 microg mL(-1)) reduced micronuclei formation by 50.7% after 1 Gy and 35.9% after 2 Gy exposures, respectively and reactive oxygen species [104]. Aloe is the most promising and widely used commercial ingredient for drugs in radiation protection and anticancer agent with good clinical efficacy. Sahebjamee *et al.*,(2015) compared the efficacy of Aloe vera mouthwash with a benzydamine mouthwash in the alleviation of radiation- induced mucositis in head and neck cancer patients using a triple-blind, randomised controlled trial [105].Twenty-six cancer patients who received conventional radiation therapy were randomised to receive an Aloe vera mouthwash or a benzydamine mouthwash. Aloe vera was found to relieve radiation induced mucositis with no side effects [105]. In one another investigation, protective effect of topical application of aloe vera gel/mild soap *versus* mild soap alone was studied in patients undergoing radiation therapy [106]. Prophylactic skin care began on the first day of radiation therapy. Patients cleansed the area with mild, unscented soap. Patients randomized into the experimental arm of the trial were instructed to liberally apply aloe vera gel to the area at various intervals throughout the day. Radioprotective effect of Aloe was found for even cumulative doses exceeding 2700cGy [106]. Aloe Vera has anticancer activity in solid neoplasm as well. In one study, including 50 patients suffering from lung cancer, gastrointestinal tract tumors, breast cancer or brain glioblastoma, were treated with pineal indole melatonin (MLT) alone (20 mg/day orally in the dark period) or MLT plus Aloe vera tincture (1 ml twice/day).Therapeutic response and disease stabilization were found in 12/24 and in 7/26 patients treated with MLT plus aloe or MLT alone, respectively [107].

CONCLUSION

Radiation induced damage has hazardous effects on health. In this review, we talked about some selected herbal drugs that can be used for intervention therapy for the treatment of radiation induced mutations and cancer. The mechanisms of radioprotection by these phytochemicals primarily rely on its antioxidant, immunomodulatory, DNA damage repair augmentation, free radical scavenging and anticancer properties. Although no drug is available till date that can revert hundred percent all the damage done by radiation to normal levels, research is needed to develop better drugs for the development of novel radioprotectors, radiomodulators, their applications, mechanisms of action, sources and chemical classifications.

CONFLICT OF INTEREST

The authors confirm that they have no conflict of interest to declare for this publication.

ACKNOWLEDGEMENT

The authors are thankful to Sharda University and NIFTEM for allowing to write this review article. The work was supported by Ramalingaswamy contingency given to me, fellowship provided to Mehak Gulzar for PhD work and Dr. Neetu K Taneja for supporting this review article.

REFERENCES

[1]   Kobashigawa S, Kashino G, Suzuki K, Yamashita S, Mori H. Ionizing radiation-induced cell death is partly caused by increase of mitochondrial reactive oxygen species in normal human fibroblast cells. Radiat Res 2015; 183(4): 455-64.
[http://dx.doi.org/10.1667/RR13772.1] [PMID: 25807320]

[2]   Little JB. Radiation carcinogenesis. Carcinogenesis 2000; 21(3): 397-404.
[http://dx.doi.org/10.1093/carcin/21.3.397] [PMID: 10688860]

[3]   Mothersill C, Seymour C. Radiation-induced bystander effects, carcinogenesis and models. Oncogene 2003; 22(45): 7028-33.
[http://dx.doi.org/10.1038/sj.onc.1206882] [PMID: 14557807]

[4]   Kumar I, Gupta D, Bhatt AN, Dwarakanath BS. Models for the development of radiation countermeasures. Def Sci J 2011; 61: 146-56.
[http://dx.doi.org/10.14429/dsj.61.835]

[5]   Carmia B, Borek C. Antioxidants and Radiation Therapy. J Nutr 2004; 134: 3207S-9S.
[PMID: 15514309]

[6]   Maurya DK, Devasagayam TP, Nair CK. Some novel approaches for radioprotection and the beneficial effect of natural products. Indian J Exp Biol 2006; 44(2): 93-114.
[PMID: 16480175]

[7]   Krishna A, Kumar A. Evaluation of radioprotective effects of Rajgira (*Amaranthus paniculatus*) extract in Swiss albino mice. J Radiat Res (Tokyo) 2005; 46(2): 233-9.
[http://dx.doi.org/10.1269/jrr.46.233] [PMID: 15988142]

[8]   Bhatia AL, Jain M. Amaranthus paniculatus (Linn.) improves learning after radiation stress. J Ethnopharmacol 2003; 85(1): 73-9.
[http://dx.doi.org/10.1016/S0378-8741(02)00337-9] [PMID: 12576205]

[9]   Maharwal J, Samarth RM, Saini MR. Radiomodulatory influence of Rajgira (*Amaranthus paniculatus*) leaf extract in Swiss albino mice. Phytother Res 2003; 17(10): 1150-4.
[http://dx.doi.org/10.1002/ptr.1340] [PMID: 14669247]

[10]   Sreelatha S, Dinesh E, Uma C. Antioxidant properties of Rajgira (Amaranthus paniculatus) leaves and potential synergy in chemoprevention. Asian Pac J Cancer Prev 2012; 13(6): 2775-80.
[http://dx.doi.org/10.7314/APJCP.2012.13.6.2775] [PMID: 22938458]

[11]   Dutta A, Lal N, Musarrat N, Ghosh A, Verma R. Ethnological and ethno-medicinal importance of aegle marmelos (l.) Corr (bael) among indigenous people of India. Am J Ethnomed 2014; 1: 290-312.

[12]   Korkina LG, Afanasev IB. Antioxidant and chelating properties of flavonoids. Adv Pharmacol 1997; 38: 151-63.
[http://dx.doi.org/10.1016/S1054-3589(08)60983-7] [PMID: 8895808]

[13]   Jagetia GC, Venkatesh P, Baliga MS. Evaluation of the radioprotective effect of *Aegle marmelos* (L.) Correa in cultured human peripheral blood lymphocytes exposed to different doses of gamma-radiation: a micronucleus study. Mutagenesis 2003; 18(4): 387-93.
[http://dx.doi.org/10.1093/mutage/geg011] [PMID: 12840113]

[14] Lampronti I, Martello D, Bianchi N, *et al. In vitro* antiproliferative effects on human tumor cell lines of extracts from the Bangladeshi medicinal plant Aegle marmelos Correa. Phytomedicine 2003; 10(4): 300-8.
[http://dx.doi.org/10.1078/094471103322004794] [PMID: 12809360]

[15] Chockalingam V, Kadali SS, Gnanasambantham P. Antiproliferative and antioxidant activity of Aegle marmelos (Linn.) leaves in Daltons Lymphoma Ascites transplanted mice. Indian J Pharmacol 2012; 44(2): 225-9.
[http://dx.doi.org/10.4103/0253-7613.93854] [PMID: 22529480]

[16] Lambertini E, Piva R, Khan MT, *et al.* Effects of extracts from Bangladeshi medicinal plants on in vitro proliferation of human breast cancer cell lines and expression of estrogen receptor alpha gene. Int J Oncol 2004; 24(2): 419-23.
[PMID: 14719119]

[17] Molls M, Beuningen D. Radiation injury of the lung: experimental studies, observations after radiotherapy and total body irradiation prior to bone marrow transplantation. In: Scherer E, Streffer C, Trott KR, Eds. Radiopathology of Organs and Tissues. Berlin: Springer-Verlag 1991; pp. 369-404.
[http://dx.doi.org/10.1007/978-3-642-83416-5_13]

[18] Xie CH, Zhang MS, Zhou YF, *et al.* Chinese medicine Angelica sinensis suppresses radiation-induced expression of TNF-alpha and TGF-beta1 in mice. Oncol Rep 2006; 15(6): 1429-36.
[PMID: 16685376]

[19] Cai HB, Luo RC. [Prevention and therapy of radiation-induced pulmonary injury with traditional Chinese medicine]. J First Mil Med Univ 2003; 23(9): 958-60.
[PMID: 13129734]

[20] Cheng TJ. [Protective action of seed oil of *Hippophae rhamnoides* L. (HR) against experimental liver injury in mice]. Zhonghua Yu Fang Yi Xue Za Zhi 1992; 26(4): 227-9.
[PMID: 1302197]

[21] Cheng TJ, Pu JK, Wu LW, Ma ZR, Cao Z, Li TJ. [An preliminary study on hepato-protective action of seed oil of Hippophae rhamnoides L. (HR) and mechanism of the action]. Zhongguo Zhong Yao Za Zhi 1994; 19(6): 367-370, 384.
[PMID: 7945887]

[22] Ianev E, Radev S, Balutsov M, Klouchek E, Popov A. [The effect of an extract of sea buckthorn (*Hippophae rhamnoides* L.) on the healing of experimental skin wounds in rats]. Khirurgiia (Sofiia) 1995; 48(3): 30-3.
[PMID: 8667579]

[23] Goel HC, Prasad J, Singh S, Sagar RK, Kumar IP, Sinha AK. Radioprotection by a herbal preparation of Hippophae rhamnoides, RH-3, against whole body lethal irradiation in mice. Phytomedicine 2002; 9(1): 15-25.
[http://dx.doi.org/10.1078/0944-7113-00077] [PMID: 11924759]

[24] Rampino M, Ricardi U, Munoz F, *et al.* Concomitant adjuvant chemoradiotherapy with weekly low-dose cisplatin for high-risk squamous cell carcinoma of the head and neck: a phase II prospective trial. Clin Oncol (R Coll Radiol) 2011; 23(2): 134-40.
[http://dx.doi.org/10.1016/j.clon.2010.09.004] [PMID: 21030225]

[25] Koeberle A, Rossi A, Bauer J, *et al.* Hyperforin, an anti-inflammatory constituent from St. Johns wort, inhibits microsomal prostaglandin E(2) synthase-1 and suppresses prostaglandin E (2) formation *in vivo.* Front Pharmacol 2011; 2: 1-10.
[http://dx.doi.org/10.3389/fphar.2011.00007] [PMID: 21779246]

[26]   Narayanan AS, Raja SS, Ponmurugan K, *et al.* Antibacterial activity of selected medicinal plants against multiple antibiotic resistant uropathogens: a study from Kolli Hills, Tamil Nadu, India. Benef Microbes 2011; 2(3): 235-43.
[http://dx.doi.org/10.3920/BM2010.0033] [PMID: 21986363]

[27]   Martínez-Poveda B, Quesada AR, Medina MA. Hyperforin, a bio-active compound of St. Johns Wort, is a new inhibitor of angiogenesis targeting several key steps of the process. Int J Cancer 2005; 117(5): 775-80.
[http://dx.doi.org/10.1002/ijc.21246] [PMID: 15981212]

[28]   dos Santos AC, Rodrigues OG, de Araojo LV, *et al.* [Use of neem extract in the control of acariasis by Myobia musculi Schranck (Acari: Miobidae) and Myocoptes musculinus Koch (Acari: Listrophoridae) in mice (Mus musculus var. albina L.)]. Neotrop Entomol 2006; 35(2): 269-72.
[PMID: 17348141]

[29]   Alam A, Haldar S, Thulasiram HV, *et al.* Novel anti-inflammatory activity of epoxyazadiradione against macrophage migration inhibitory factor: inhibition of tautomerase and proinflammatory activities of macrophage migration inhibitory factor. J Biol Chem 2012; 287(29): 24844-61.
[http://dx.doi.org/10.1074/jbc.M112.341321] [PMID: 22645149]

[30]   Franco1 P, Potenza I, Moretto F, Segantin M, Grosso M, *et al.* Hypericum perforatum and neem oil for the Management of acute skin toxicity in head and neck cancer patients undergoing radiation or Chemo-radiation: a single-arm prospective observational study. Radiat Oncol 2014; 9: 297-302.
[http://dx.doi.org/10.1186/s13014-014-0297-0] [PMCID: PMC4300176]

[31]   Kitts D, Hu C. Efficacy and safety of ginseng. Public Health Nutr 2000; 3(4A): 473-85.
[http://dx.doi.org/10.1017/S1368980000000550] [PMID: 11276295]

[32]   Yun TK. Brief introduction of Panax ginseng C.A. Meyer. J Korean Med Sci 2001; 16 (Suppl.): S3-5.
[http://dx.doi.org/10.3346/jkms.2001.16.S.S3] [PMID: 11748372]

[33]   Shibata S. Chemistry and cancer preventing activities of ginseng saponins and some related triterpenoid compounds. J Korean Med Sci 2001; 16 (Suppl.): S28-37.
[http://dx.doi.org/10.3346/jkms.2001.16.S.S28] [PMID: 11748374]

[34]   Caron MF, Hotsko AL, Robertson S, Mandybur L, Kluger J, White CM. Electrocardiographic and hemodynamic effects of Panax ginseng. Ann Pharmacother 2002; 36(5): 758-63.
[http://dx.doi.org/10.1345/aph.1A411] [PMID: 11978148]

[35]   Yun TK, Lee YS, Lee YH, Kim SI, Yun HY. Anticarcinogenic effect of Panax ginseng C.A. Meyer and identification of active compounds. J Korean Med Sci 2001; 16 (Suppl.): S6-S18.
[http://dx.doi.org/10.3346/jkms.2001.16.S.S6] [PMID: 11748383]

[36]   Yun TK. Experimental and epidemiological evidence on non-organ specific cancer preventive effect of Korean ginseng and identification of active compounds. Mutat Res 2003; 523-524: 63-74.
[http://dx.doi.org/10.1016/S0027-5107(02)00322-6] [PMID: 12628504]

[37]   Attele AS, Wu JA, Yuan CS. Ginseng pharmacology: multiple constituents and multiple actions. Biochem Pharmacol 1999; 58(11): 1685-93.
[http://dx.doi.org/10.1016/S0006-2952(99)00212-9] [PMID: 10571242]

[38]   Gillis CN. Panax ginseng pharmacology: a nitric oxide link? Biochem Pharmacol 1997; 54(1): 1-8.
[http://dx.doi.org/10.1016/S0006-2952(97)00193-7] [PMID: 9296344]

[39]   Song JY, Han SK, Bae KG, *et al.* Radioprotective effects of ginsan, an immunomodulator. Radiat Res 2003; 159(6): 768-74.
[http://dx.doi.org/10.1667/0033-7587(2003)159[0768:REOGAI]2.0.CO;2] [PMID: 12751959]

[40]   Ben-Hur E, Fulder S. Effect of Panax ginseng saponins and Eleutherococcus senticosus on survival of cultured mammalian cells after ionizing radiation. Am J Chin Med 1981; 9(1): 48-56.

[http://dx.doi.org/10.1142/S0192415X8100007X] [PMID: 7304498]

[41] Kim TH, Lee YS, Cho CK, Park S, Choi SY, Yool SY. Protective effect of ginseng on radiation-induced DNA double strand breaks and repair in murine lymphocytes. Cancer Biother Radiopharm 1996; 11(4): 267-72.
[http://dx.doi.org/10.1089/cbr.1996.11.267] [PMID: 10851547]

[42] Kim SH, Jeong KS, Ryu SY, Kim TH. Panax ginseng prevents apoptosis in hair follicles and accelerates recovery of hair medullary cells in irradiated mice. In Vivo 1998; 12(2): 219-22.
[PMID: 9627805]

[43] Hsu HY, Yang JJ, Lian SL, Ho YH, Lin CC. Recovery of the hematopoietic system by Si-Jun-Zi-Tang in whole body irradiated mice. J Ethnopharmacol 1996; 54(2-3): 69-75.
[http://dx.doi.org/10.1016/S0378-8741(96)01450-X] [PMID: 8953420]

[44] Kumar M, Sharma MK, Saxena PS, Kumar A. Radioprotective effect of Panax ginseng on the phosphatases and lipid peroxidation level in testes of Swiss albino mice. Biol Pharm Bull 2003; 26(3): 308-12.
[http://dx.doi.org/10.1248/bpb.26.308] [PMID: 12612438]

[45] Makkar HP, Becker K. Nutritional value and antinutritional components of whole and ethanol extracted Moringa oleifera leaves. Anim Feed Sci Technol 1996; 63: 211-28.
[http://dx.doi.org/10.1016/S0377-8401(96)01023-1]

[46] Selvakumar D, Natarajan P. Hepatoprotective activity of Moringaoleifera Lam in carbon tetrachloride induced hepatotoxicity in albino rats. Phcog Mag 2008; 4: 97-8.

[47] Fakurazi S, Hairuszah I, Nanthini U. Moringa oleifera Lam prevents acetaminophen induced liver injury through restoration of glutathione level. Food Chem Toxicol 2008; 46(8): 2611-5.
[http://dx.doi.org/10.1016/j.fct.2008.04.018] [PMID: 18514995]

[48] Rao AV, Devi PU, Kamath R. *In vivo* radioprotective effect of Moringa oleifera leaves. Indian J Exp Biol 2001; 39(9): 858-63.
[PMID: 11831365]

[49] Sinha M, Das DK, Bhattacharjee S, Majumdar S, Dey S. Leaf extract of Moringa oleifera prevents ionizing radiation-induced oxidative stress in mice. J Med Food 2011; 14(10): 1167-72.
[http://dx.doi.org/10.1089/jmf.2010.1506] [PMID: 21861723]

[50] Jirovetz L, Buchbauer G, Wobus A, *et al.* Medicinally used plants from India:Analysis of the essential oil of air dried *Biophytum sensitivum* (L.) DC. Sci Pharm 2004; 72: 87-96.

[51] Inngjerdingen KT, Coulibaly A, Diallo D, Michaelsen TE, Paulsen BS. A complement fixing polysaccharide from *Biophytum petersianum* Klotzsch, a medicinal plant from Mali, West Africa. Biomacromolecules 2006; 7(1): 48-53.
[http://dx.doi.org/10.1021/bm050330h] [PMID: 16398497]

[52] Jachak SM, Bucar F, Kartnig T. Antiinflammatory activity of extracts of *Biophytum sensitivum* in carrageenin-induced rat paw oedema. Phytother Res 1999; 13(1): 73-4.
[http://dx.doi.org/10.1002/(SICI)1099-1573(199902)13:1<73::AID-PTR374>3.0.CO;2-V] [PMID: 10189957]

[53] Puri D. The insulinotropic activity of a Nepalese medicinal plant *Biophytum sensitivum*: preliminary experimental study. J Ethnopharmacol 2001; 78(1): 89-93.
[http://dx.doi.org/10.1016/S0378-8741(01)00306-3] [PMID: 11585694]

[54] Guruvayoorappan C, Afira AH, Kuttan G. Antioxidant potential of *Biophytum sensitivum* extract *in vitro* and *in vivo*. J Basic Clin Physiol Pharmacol 2006; 17(4): 255-67.
[http://dx.doi.org/10.1515/JBCPP.2006.17.4.255] [PMID: 17338281]

[55]   Guruvayoorappan C, Kuttan G. Anti-angiogenic effect of *Biophytum sensitivum* is exerted through its cytokine modulation activity and inhibitory activity againstVEGF mRNA expression, endothelial cell migration and capillary tube formation. J Exp Ther Oncol 2007; 6: 241-50.
[PMID: 17552364]

[56]   Guruvayoorappan C, Kuttan G. Immunomodulatory and antitumor activity of *Biophytum sensitivum* extract. Asian Pac J Cancer Prev 2007; 8(1): 27-32.
[PMID: 17477767]

[57]   Guruvayoorappan C, Kuttan G. Protective effect of Biophytum sensitivum (L.) DC on radiation-induced damage in mice. Immunopharmacol Immunotoxicol 2008; 30(4): 815-35.
[http://dx.doi.org/10.1080/08923970802439480] [PMID: 18951225]

[58]   Huang L, Zhao H, Huang B, Zheng C, Peng W, Qin L. Acanthopanax senticosus: review of botany, chemistry and pharmacology. Pharmazie 2011; 66(2): 83-97. [Review].
[PMID: 21434569]

[59]   Miyanomae T, Frindel E. Radioprotection of hemopoiesis conferred by Acanthopanax senticosus Harms (Shigoka) administered before or after irradiation. Exp Hematol 1988; 16(9): 801-6.
[PMID: 3049132]

[60]   Bespalov VG, Alexandrov VA, Semenov AL, KovanKo EG, Ivanov SD. Anticarcinogenic activity of alpha-difluoromethylornithine, ginseng, eleutherococcus, and leuzea on radiation-induced carcinogenesis in female rats. Int J Radiat Biol 2014; 90(12): 1191-200. [Review].
[http://dx.doi.org/10.3109/09553002.2014.932937] [PMID: 24913295]

[61]   Bałan BJ, Niemcewicz M, Kocik J, Jung L, Skopińska-Różewska E, Skopiński P. Oral administration of Aloe vera gel, anti-microbial and anti-inflammatory herbal remedy, stimulates cell-mediated immunity and antibody production in a mouse model. Cent Eur J Immunol 2014; 39(2): 125-30.
[http://dx.doi.org/10.5114/ceji.2014.43711] [PMID: 26155113]

[62]   Sahu PK, Giri DD, Singh R, Pandey P, *et al.* Therapeutic and medicinal uses of Aloe vera: a review. Pharmacol Pharm 2013; 4: 599-610.
[http://dx.doi.org/10.4236/pp.2013.48086]

[63]   Saini DK, Saini MR. Evaluation of radioprotective efficacy and possible mechanism of action of Aloe gel. Environ Toxicol Pharmacol 2011; 31(3): 427-35.
[http://dx.doi.org/10.1016/j.etap.2011.02.004] [PMID: 21787713]

[64]   Byeon SW, Pelley RP, Ullrich SE, Waller TA, Bucana CD, Strickland FM. Aloe barbadensis extracts reduce the production of interleukin-10 after exposure to ultraviolet radiation. J Invest Dermatol 1998; 110(5): 811-7.
[http://dx.doi.org/10.1046/j.1523-1747.1998.00181.x] [PMID: 9579551]

[65]   Saada HN, Ussama ZS, Mahdy AM. Effectiveness of Aloe vera on the antioxidant status of different tissues in irradiated rats. Pharmazie 2003; 58(12): 929-31.
[PMID: 14703976]

[66]   de Witte P. Metabolism and Pharmacokinetics of An-thranoids. Pharmacology 1993; 47: 86-97.

[67]   Syed TA, Afzal M, Ashfaq AS. Management of Genital Herpe in men with 0.5% *Aloe vera* extracts in a hydrophilic cream: A placebo-controlled double-blind study. J Dermatolog Treat 1997; 8: 99-102.
[http://dx.doi.org/10.3109/09546639709160279]

[68]   Visuthikosol V, Chowchuen B, Sukwanarat Y, Sriurairatana S, Boonpucknavig V. Effect of *aloe vera* gel to healing of burn wound a clinical and histologic study. J Med Assoc Thai 1995; 78(8): 403-9.
[PMID: 7561562]

[69]   Fulton JE Jr. The stimulation of postdermabrasion wound healing with stabilized aloe vera gel-polyethylene oxide dressing. J Dermatol Surg Oncol 1990; 16(5): 460-7.
[http://dx.doi.org/10.1111/j.1524-4725.1990.tb00065.x] [PMID: 2341661]

[70]   Rabe T, van Staden J. Antibacterial activity of South African plants used for medicinal purposes. J Ethnopharmacol 1997; 56(1): 81-7.
[http://dx.doi.org/10.1016/S0378-8741(96)01515-2] [PMID: 9147258]

[71]   Joseph B, Raj SJ. Pharmacognostic and Phyto-chemical Properties of *Aloe vera* Linn—an overview. Int J Pharm Sci Rev Res 2010; 4: 106-10.

[72]   Bräuchler C, Meimberg H, Heubl G. Molecular phylogeny of Menthinae (Lamiaceae, Nepetoideae, Mentheae)Taxonomy, biogeography and conflicts. Mol Phylogenet Evol 2010; 55(2): 501-23.
[http://dx.doi.org/10.1016/j.ympev.2010.01.016] [PMID: 20152913]

[73]   Baliga MS, Rao S. Radioprotective potential of mint: a brief review. J Cancer Res Ther 2010; 6(3): 255-62.
[http://dx.doi.org/10.4103/0973-1482.73336] [PMID: 21119249]

[74]   Jagetia GC, Baliga MS. Influence of the leaf extract of *Mentha arvensis Linn.* (mint) on the survival of mice exposed to different doses of gamma radiation. Strahlenther Onkol 2002; 178(2): 91-8.
[http://dx.doi.org/10.1007/s00066-002-0841-y] [PMID: 11942043]

[75]   Samarth RM, Saini MR, Maharwal J, Dhaka A, Kumar A. Mentha piperita (Linn) leaf extract provides protection against radiation induced alterations in intestinal mucosa of Swiss albino mice. Indian J Exp Biol 2002; 40(11): 1245-9.
[PMID: 13677626]

[76]   Hassan HA, Hafez HS, Goda MS. Mentha piperita as a pivotal neuro-protective agent against gamma irradiation induced DNA fragmentation and apoptosis : Mentha extract as a neuroprotective against gamma irradiation. Cytotechnology 2013; 65(1): 145-56.
[http://dx.doi.org/10.1007/s10616-012-9470-1] [PMID: 23011739]

[77]   Samarth RM, Panwar M, Kumar M, Kumar A. Radioprotective influence of Mentha piperita (Linn) against gamma irradiation in mice: Antioxidant and radical scavenging activity. Int J Radiat Biol 2006; 82(5): 331-7.
[http://dx.doi.org/10.1080/09553000600771523] [PMID: 16782650]

[78]   Mishra B, Priyadarsini KI, Kumar MS, Unnikrishnan MK, Mohan H. Effect of O-glycosilation on the antioxidant activity and free radical reactions of a plant flavonoid, chrysoeriol. Bioorg Med Chem 2003; 11(13): 2677-85.
[http://dx.doi.org/10.1016/S0968-0896(03)00232-3] [PMID: 12788341]

[79]   Prabhakar KR, Veerapur VP, Parihar KV, Priyadarsini KI, Rao BS, Unnikrishnan MK. Evaluation and optimization of radioprotective activity of Coronopus didymus Linn. in gamma-irradiated mice. Int J Radiat Biol 2006; 82(8): 525-36.
[http://dx.doi.org/10.1080/09553000600876686] [PMID: 16966180]

[80]   Chopra RN, Nayar SL, Chopra IC. Glossary of Indian medicinal plants (Including the Supplement). New Delhi: Council of Scientific and Industrial Research 1986; pp. 51-83.

[81]   Diwan PC, Karwande I, Singh AK. Anti-anxiety profile of mandukparni *Centella asiatica* Linn in animals. Fitoterapia 1991; 62: 255-7.

[82]   Shobi V, Goel HC. Protection against radiation-induced conditioned taste aversion by *Centella asiatica*. Physiol Behav 2001; 73(1-2): 19-23.
[http://dx.doi.org/10.1016/S0031-9384(01)00434-6] [PMID: 11399290]

[83]   Joy J, Nair CK. Protection of DNA and membranes from gamma-radiation induced damages by Centella asiatica. J Pharm Pharmacol 2009; 61(7): 941-7.
[http://dx.doi.org/10.1211/jpp.61.07.0014] [PMID: 19589237]

[84]   An IS, An S, Kang SM, *et al.* Titrated extract of Centella asiatica provides a UVB protective effect by altering microRNA expression profiles in human dermal fibroblasts. Int J Mol Med 2012; 30(5): 1194-202.
[PMID: 22948173]

[85]   Yokota T, Nishio H, Kubota Y, Mizoguchi M. The inhibitory effect of glabridin from licorice extracts on melanogenesis and inflammation. Pigment Cell Res 1998; 11(6): 355-61.
[http://dx.doi.org/10.1111/j.1600-0749.1998.tb00494.x] [PMID: 9870547]

[86]   Das D, Agarwal SK, Chandola HM. Protective effect of Yashtimadhu (Glycyrrhiza glabra) against side effects of radiation/chemotherapy in head and neck malignancies. Ayu 2011; 32(2): 196-9.
[http://dx.doi.org/10.4103/0974-8520.92579] [PMID: 22408302]

[87]   Song NR, Kim JE, Park JS, *et al.* Licochalcone A, a polyphenol present in licorice, suppresses UV-induced COX-2 expression by targeting PI3K, MEK1, and B-Raf. Int J Mol Sci 2015; 16(3): 4453-70.
[http://dx.doi.org/10.3390/ijms16034453] [PMID: 25710724]

[88]   Palbag S, Dey BK, Singh NK. Ethnopharmacology, phytochemistry and pharmacology of Tephrosia purpurea. Chin J Nat Med 2014; 12(1): 1-7.
[http://dx.doi.org/10.1016/S1875-5364(14)60001-7] [PMID: 24484589]

[89]   Luo ZP, Lin HY, Ding WB, He HL, Li YZ. Phylogenetic diversity and antifungal activity of endophytic fungi associated with Tephrosia purpurea. Mycobiology 2015; 43(4): 435-43.
[http://dx.doi.org/10.5941/MYCO.2015.43.4.435] [PMID: 26839503]

[90]   Nile SH, Park SW. HPTLC analysis, antioxidant and antigout activity of Indian plants. Iran J Pharm Res 2014; 13(2): 531-9.
[PMID: 25237348]

[91]   Palbag S, Dey BK, Singh NK. Ethnopharmacology, phytochemistry and pharmacology of Tephrosia purpurea. Chin J Nat Med 2014; 12(1): 1-7.
[http://dx.doi.org/10.1016/S1875-5364(14)60001-7] [PMID: 24484589]

[92]   Patel A, Patel A, Patel A, Patel NM. Determination of polyphenols and free radical scavenging activity of Tephrosia purpurea linn leaves (Leguminosae). Pharmacognosy Res 2010; 2(3): 152-8.
[http://dx.doi.org/10.4103/0974-8490.65509] [PMID: 21808558]

[93]   Choudhary GP. *In vitro* antioxidant studies of the ethanolic extract of Tephrosia purpurea L. Anc Sci Life 2007; 27(1): 26-30.
[PMID: 22557256]

[94]   Yang B, Liu P. Composition and biological activities of hydrolyzable tannins of fruits of Phyllanthus emblica. J Agric Food Chem 2014; 62(3): 529-41.
[http://dx.doi.org/10.1021/jf404703k] [PMID: 24369850]

[95]   Majeed M, Bhat B, Anand S, Sivakumar A, Paliwal P, Geetha KG. Inhibition of UV-induced ROS and collagen damage by Phyllanthus emblica extract in normal human dermal fibroblasts. J Cosmet Sci 2011; 62(1): 49-56.
[PMID: 21443845]

[96]   Adil MD, Kaiser P, Satti NK, Zargar AM, Vishwakarma RA, Tasduq SA. Effect of Emblica officinalis (fruit) against UVB-induced photo-aging in human skin fibroblasts. J Ethnopharmacol 2010; 132(1): 109-14.
[http://dx.doi.org/10.1016/j.jep.2010.07.047] [PMID: 20688142]

[97]   Jindal A, Soyal D, Sharma A, Goyal PK. Protective effect of an extract of Emblica officinalis against radiation-induced damage in mice. Integr Cancer Ther 2009; 8(1): 98-105.
[http://dx.doi.org/10.1177/1534735409331455] [PMID: 19223372]

[98]   Singh I, Soyal D, Goyal PK. Emblica officinalis (Linn.) fruit extract provides protection against radiation-induced hematological and biochemical alterations in mice. J Environ Pathol Toxicol Oncol 2006; 25(4): 643-54.
[http://dx.doi.org/10.1615/JEnvironPatholToxicolOncol.v25.i4.40] [PMID: 17341205]

[99]   Singh I, Sharma A, Nunia V, Goyal PK. Radioprotection of Swiss albino mice by Emblica officinalis. Phytother Res 2005; 19(5): 444-6.
[http://dx.doi.org/10.1002/ptr.1600] [PMID: 16106381]

[100]   Sancheti G, Jindal A, Kumari R, Goyal PK. Chemopreventive action of emblica officinalis on skin carcinogenesis in mice. Asian Pac J Cancer Prev 2005; 6(2): 197-201.
[PMID: 16101333]

[101]   De A, De A, Papasian C, *et al.* Emblica officinalis extract induces autophagy and inhibits human ovarian cancer cell proliferation, angiogenesis, growth of mouse xenograft tumors. PLoS One 2013; 8(8): e72748.
[http://dx.doi.org/10.1371/journal.pone.0072748] [PMID: 24133573]

[102]   Kupin VI, Polevaia EB. [Stimulation of the immunological reactivity of cancer patients by Eleutherococcus extract]. Vopr Onkol 1986; 32(7): 21-6. [Russian.].
[PMID: 3526720]

[103]   Zhuang SR, Chiu HF, Chen SL, *et al.* Effects of a Chinese medical herbs complex on cellular immunity and toxicity-related conditions of breast cancer patients. Br J Nutr 2012; 107(5): 712-8.
[http://dx.doi.org/10.1017/S000711451100345X] [PMID: 21864416]

[104]   Lee TK, OBrien KF, Wang W, *et al.* Radioprotective effect of American ginseng on human lymphocytes at 90 minutes postirradiation: a study of 40 cases. J Altern Complement Med 2010; 16(5): 561-7.
[http://dx.doi.org/10.1089/acm.2009.0590] [PMID: 20491513]

[105]   Sahebjamee M, Mansourian A, Hajimirzamohammad M, *et al.* Comparative efficacy of Aloe vera and Benzydamine Mouthwashes on Radiation-induced Oral Mucositis: A triple-blind, randomised, controlled clinical trial. Oral Health Prev Dent 2015; 13(4): 309-15.
[PMID: 25431805]

[106]   Olsen DL, Raub W Jr, Bradley C, *et al.* The effect of aloe vera gel/mild soap *versus* mild soap alone in preventing skin reactions in patients undergoing radiation therapy. Oncol Nurs Forum 2001; 28(3): 543-7.
[PMID: 11338761]

[107]   Lissoni P, Giani L, Zerbini S, Trabattoni P, Rovelli F. Biotherapy with the pineal immunomodulating hormone melatonin *versus* melatonin plus aloe vera in untreatable advanced solid neoplasms. Nat Immun 1998; 16(1): 27-33.
[http://dx.doi.org/10.1159/000069427] [PMID: 9789122]

CHAPTER 7

Marine Natural Products for Cancer Prevention and Therapy: A Mechanistic Overview

Shankar Suman, Sanjay Mishra and **Yogeshwer Shukla**[*]

Proteomics and Environmental Carcinogenesis Laboratory, CSIR-Indian Institute of Toxicology Research, M.G. Marg, Lucknow, India

Abstract: Marine resources have rich pharmaceutical values as they encompass a diverse taxonomy of biological species and possess a large-scale of bioactive compounds. The existence of extraordinary chemical diversity of marine resources is used to discover anticancer agent in its natural or derived synthetic form. Marine flora and fauna possess several valuable compounds with immunostimulatory and antioxidant properties, also gained their importance as nutraceuticals as well as in cancer chemoprevention. In this chapter, we have emphasized for overviewing the marine natural products; those having anticancer or cancer chemo-preventive properties, acting against different deregulated cellular and molecular pathways associated with cancer development or progression.

Keywords: Cancer, Cell signaling pathways, Flora, Fauna, Marine compounds.

INTRODUCTION

Cancer is a global health concern with a high rate of mortality and morbidity. In the recent GLOBOCAN report, 1 million new cases and 8.2 million deaths due to cancer were reported in 2012 [1]. The incidence of cancer has been increasing drastically in the past few decades. Epidemiological studies on cancer showed that various environmental factors and irregular lifestyle are key governing elements of cancer development. Thus, several of naturally available preventive agents are well explored to manage cancer incidence and progression. Marine resources have enormous biological diversity and possess a huge content of bioactive compounds. Hence, there is a growing interest on the screening of marine compounds that possess potent cancer preventive abilities *via* targeting various cellular and molecular machineries associated with cancer. So far, a large-scale

[*] **Corresponding author Yogeshwer Shukla**: Proteomics and Environmental Carcinogenesis Laboratory, CSIR-Indian Institute of Toxicology Research, M.G. Marg, Lucknow, India; Tel/Fax: (0091) 522-2628227; E-mail: yshukla@iitr.res.in

Sahdeo Prasad & Amit Kumar Tyagi (Eds.)

natural compounds or its derived form of bioactive-compounds are widely studied for their anticancer activities in numerous studies. These compounds have unique abilities to target cell signaling as well as metabolic pathways which are involved in cancer associated cellular processes. Several marine organisms produce a variety of secondary metabolites, which are known for the cancer prevention activity, such as polyphenols, alkaloids, porphyrins, terpenoids, phenazines, fatty acid products, sterols, amino acid analogues, aliphatic cyclic peroxides, peptides *etc*. In the present chapter, several cancer preventive agents are elaborated with their potent targets in cancer-associated pathways, which have a major role in the cancer development and progression.

How Cancer Preventive or Therapeutic Agents Alter Carcinogenesis?

In the last few decades, multiple views from the investigations have delineated the mechanistic pathways in the carcinogenesis. The present conception of neoplastic transformations is the DNA damage and accumulating mutations at genomic level are one of the key factors of the carcinogenic process and these are highly integrated with various types of cellular stresses. Several factors, including hectic lifestyle and environmental toxicants generated cellular stress in the oxidative forms by the metabolic process, which become one of the reasons to disrupt cell machinery by lipid peroxidation, DNA adduct formation, *etc*. and ultimately lead to an oncogenic event. However, several mutations can be overcome by repair mechanism, thus the defective repair system also supports the oncogenic process. Multiple molecular mechanisms are associated with the initiation and progression of different human malignancies, which are centrally linked to the physiological regulators of homeostasis. Thus, activation of cell survival factors like AKT, MAPK and inhibition of apoptosis are known to assist in the proliferation of cancer cells. Several investigations on cancer suggested that the factors connecting link of cell survival to cell death in cancer could be critically important for therapeutic intervention in cancer. Presently, many of the marine natural products are well known for their role in intervening the cancer-associated molecular mechanism; these are directly or indirectly linked to deregulated metabolic pathways or physiological process in cancer progression. The majority of isolated compounds from marine source are peptides, lipids, metabolites *etc*. which are vitally reported for their anticancer activities. Due to their unique anticancer potential, these products also become a choice for the pharmaceutical industry to discover anticancer drugs. From the recent opinions of cancer-associated mechanisms and their targets, marine products emerged as targets for various mechanistic pathways to diminish cancer progression. These compounds target various major pathways involved in the cancer progression, which are detailed in the subheadings (Fig. **1**).

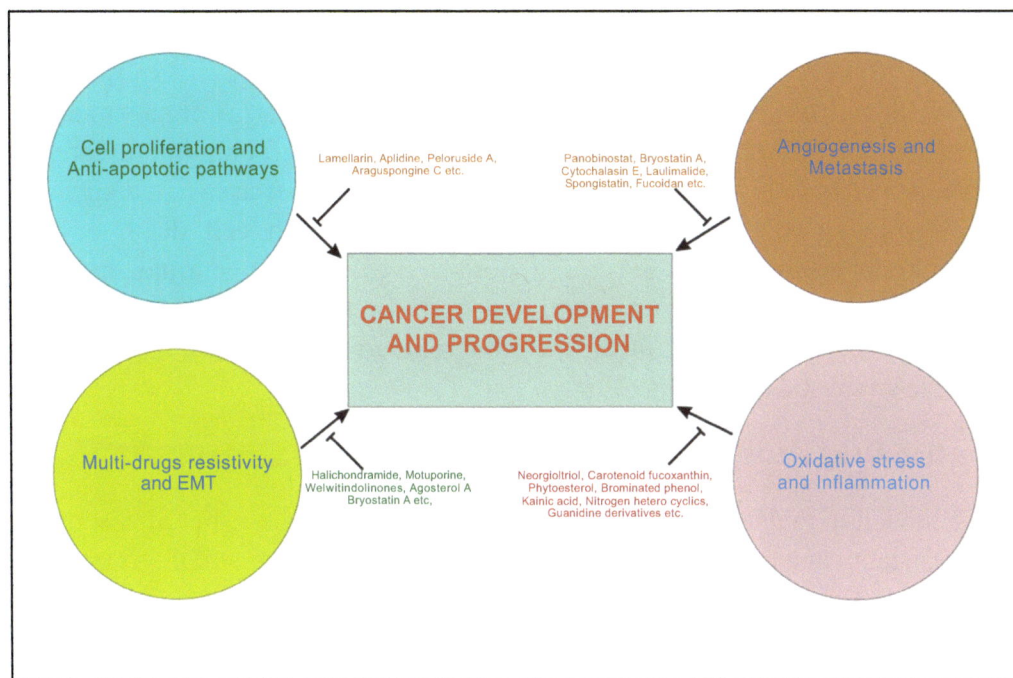

Fig. (1). Current views of the mechanism associated with cancer development and progression. The large-scale marine natural compounds are having unique potency to target these cancer associated pathways. In the figure, few marine compounds are mentioned, which prevent these deregulated processes associated with cancer development and progression.

Targeting Cancer Associated Signaling Pathways

Natural compounds from marine flora and fauna have been explored for their anticancer properties with multiple molecular targets of cancer signaling pathways. It is widely known that cancer cells can grow under constitutive activation of proto-oncogene to oncogenes and or inactivation of tumor suppressor gene with genetic abnormalities. These activation/suppression of various oncogene and suppressor genes followed a multiple signaling pathway, leading to an initiation and progression of cancer. A large number of investigations have demonstrated the efficacy of natural products to control these altered pathways in cancer. As many deregulated pathways implicated in cancer, most commonly defective p53 possess a crosstalk of multiple signaling pathways associated with the proliferation of cancer cell. Several compounds from marine sources exert antiproliferative effects *via* p53 activation. Marine sponge's metabolites (ilimaquinone and ethylsmenoquinone) as activators of p53 pathway have shown to induce apoptosis and autophagy in cancer cells [2]. PI3K/Akt signaling pathway is another target of cancer, as it is frequently over-expressed in several types of cancer. Recently, it is observed that marine lipopeptide, such as Iturin A can trigger apoptosis in breast cancer cells by inhibiting Akt mediated signaling [3]. TNF-related apoptosis-inducing ligand (TRAIL) is known to be selective agents for treating cancer, aplycin derived from marine organism acts as a sensitizer for TRAIL by p38 MAPK/survivin [4]. It is still ongoing research on marine products to investigate the various targets to cancer associated signaling pathways that are directly or indirectly associated with cell death or anti-proliferative mechanisms. Several reports revealed that marine bioactive compounds also inhibit onco-preventive activity by modulating various cellular and molecular pathways in cancer, including oxidative stress, anti-angiogenesis, cell cycle arrest and cell death.

Targeting Drug Resistance Transporter Proteins

These bioactive molecules not only target signaling pathways, but also sensitize drug resistant cancer cells by targeting multi drug resistance (MDR) transporter proteins, such as P-glycoprotein, breast cancer resistance protein (BCRP), and multidrug resistance protein 1 (MRP1), belonging to the ATP-binding cassette (ABC) transporter proteins. These proteins also play a key role in drug efflux from cancer cells through which cancer cells survive after chemotherapy. Targeting cancer cells *via* p-glycoproteins is one of the important options for drug sensitivity and cancer therapy. Hence, the several inhibitors of P-glycoproteins are in clinical trials. Several of the P-glycoproteins inhibitors are currently known, belonging to a variety of bioactive compounds from marine resources (Table **1**). The approved marine natural products and their analogues such as vidarbine, trabectedin, cytarabine, ziconotide and halaven are in clinical use [5]. Lopez *et al* recently reviewed the list of marine products targeting P-glycoprotein. These products include Sipholenol A, Lamellarin, Agosterol A, ET-743, N-Met--ylwelwitindolinone-C-Isothiocyanate, Parguerenes, Kendarimide A, Bryostatin 1, ISA, Nocardioazines, Discodermolide and Polyoxygenated steroids [6]. Major of these drugs are reported to enhance chemo-sensitivity of several routinely used chemotherapeutic drugs for different types of cancers. BCRP, MRP1 and other major transporters are over-expressed in a number of cancer types. A list of marine products act as inhibitors against these types of transporters and a few of them are enlisted in Table **1**.

Table 1. Marine products act against drug resistance transporters.

Sr. no.	Name	Source	Action on MDR	References
1.	Lamellarin (Pyrrole Alkaloid)	*Australian marine sponge, Lanthella sp.,*	Reverses BCRP in Cancer Cells	[7]
2.	Welwitindolinones	*Hapalosiphon welwitschii* Blue green alga	Reverses P-glycoprotein MDR	[8]
3.	Agosterol A (AG-A)	*Spongia* sp., Sponges	Reverses P-glycoprotein and MRP1	[9, 10]
4.	Sipholenol A	*Callyspongia siphonella*	Reverses P-glycoprotein	[11]
5.	Kendarimide A	*Haliclona sp.* (marine Sponge)	Novel peptide reversing P-Glycoprotein	[12]
6.	Shornephine A (1)	Marine-derived *Aspergillus sp.*	Non-cytotoxic inhibitor of P-Glycoprotein	[13]
7.	Bryostatin 1 macrocyclic lactone	marine invertebrate *Bugula neritina*	Reverses P-glycoprotein	[14]

Inducing Cell Cycle Arrest, Cell Death and Immunomodulation

Various targeting strategies are used to inhibit progression of cancer cells, in which blockade of cell cycle progression is one of the important strategies for cancer treatment. Cancer cell proliferation can be inhibited by targeting cell arrest at G1, G2 or M phase of cell cycle through modulating regulatory proteins of different phases of the cell cycle. Aplidine, from mediterranean tunicate (*Aplidium albicans*) is currently in phase II clinical trial known to induce both G1 and G2 arrest [15]. Higher cell proliferation and mal-functioning the cell death are common features of tumor cells survival and progression. Marine natural compounds have many targets to inhibit tumor growth by altering cellular and molecular machineries of cancer cells. Any abnormal cell can undergo into cell death by any of the process like apoptosis, autophagy, senescence and necrosis. Apoptosis is understood as an unbiased or natural cell death process. A large scale of marine compounds showed anti-apoptotic activities and cancer cell death, such as marine derived Steroid methyl spongoate extracted from a soft coral *Spongodes* sp. showed the cytotoxicity in liver cancer cells *via* apoptosis induction [16]. Autophagy has dual roles during the process of cancer suppression or progression. At one side, it acts to degrade damaged cells and prevents proteins and organelle accumulation and play a fundamental role for tumor suppression [17]. In many cancer types, due to high metabolic demand and cellular stress, tumor cells promote autophagy to support a rapid cell proliferation and tumor development [17]. However, it is also reported that Araguspongine C (*Xestospongia* sp.)

induces autophagy death in breast cancer cells [18]. Several other marine compounds also act as inhibitors or inducers of autophagy, and both could be used as potential preventive agents for cancer. Necrosis is acute or accidental cell death process, causing inflammation and failing to successfully eliminate cancer cells. It can also be generated due to cytotoxic drugs and one of the reasons of side effect of chemotherapeutic drug. With a novel combination of marine natural products and chemotherapeutic drugs, cancer cells may be chemo sensitized to apoptosis and this strategy also limits the undesired toxicity as caused by chemotherapeutic drugs. Immunomodulation is also a very important feature of marine compounds, because these compounds are also able to activate several phagocytic immune cells as well as cytokines and interleukins. Fucoidan is polysaccharide, extracted from brown seaweed showing a maturing effect on dendritic cells [19]. Polysaccharides isolated from marine fungus *Phoma herbarum* exhibited specific immunomodulatory effects by modulating dendritic cells (DCs) and T cells [20].

Target Different Cellular Machinery to Suppress Tumor Development

Several marine compounds have been found to possess an exceptional potency to target actin and microtubule for inhibiting cancer cell proliferation. Several investigations also added a crucial role of microtubule in the regulation of endocrine signaling pathways in cancer [21]. Peloruside A, isolated compound from marine sponge *Mycale* acts as a novel microtubule-stabilizing agent and possess anticancer activities [22]. JG6, a novel compound from marine source act to bind at actin-binding sites of cofilin (known to dissemble actin filament) and hence, it plays significant role in treating metastatic cancer [23]. Matrix metalloproteinase imparts a role in the degradation of extracellular matrix and tissue remodeling, which is imbalanced in many chronic diseases including cancer [24]. Metalloproteinases inhibitors (MMPIs) are now increasingly used for cancer management; several of them are also used in clinical trials. The unparalleled contribution of marine natural products propounded their applications as various MMPIs; most of them extracted from seaweeds [24]. Various compounds from marine extracts have targeted angiogenesis (a common features occurring in tumor development), a common marine derived angiogenic inhibitor is panobinostat, the most potent histone deacetylases, which is under clinical trials [25].

Major Biological Resources of Cancer Preventive Natural Marine Products

A variety of marine natural compounds has an array of applications for cancer management taking from prevention to treatment. A large number of bioactive compounds from marine flora and fauna have propounded the anticancer activities, and some of them are under clinical trials (https://www.clinical-trials.gov) (Table **2**). Multiple signaling pathways involved in carcinogenesis can be targeted by various isolated compounds from marine biological resources and hence, these are promising sources of drug discovery for cancer. The details of anticancer compounds and their biological sources have been given in the subsections.

Table 2. List of anti-cancer marine compounds in different clinical trials (Phase II afterwards).

Source organisms	Compounds	Possible targets	Status with Clinical trial identifier[#]
Dolabella auricularia/ Symploca sp. *(Mollusc/cyanobacterium)*	Dolastatin 10 (linear peptide)	Tubulin	Phase II NCT00003778 NCT00003677 NCT00003557 NCT00003914 NCT00005579 NCT00003693 NCT00003626
Bugula neritina (bryozoan)	Bryostatin 1 (Macrocyclic lactone)	PKC	Phase II NCT00031694 NCT00058305 NCT00032188 NCT00006942 NCT00006389 NCT00136461 NCT00002725 NCT00003968 NCT00005849 NCT00005028 NCT00087425 Phase I NCT00112476 NCT00003166 NCT00012376 NCT00004144
Dolabella auricularia/ Symploca sp. (synthetic analogue)	Synthadotin (Linear peptide)	Tubulin	Phase II NCT00082134 NCT00078455 NCT00068211
Elysia rufescens (mollusk)	Kahalalide F (Cyclic depsipeptide)	Lysosomes/erbB Pathway	Phase II EudraCT No. 2004-001253-29
Squalus acanthias (shark)	Squalamine	Phospholipid bilayer	Phase II NCT00021385

(Table 2) contd.....

Source organisms	Compounds	Possible targets	Status with Clinical trial identifier[#]
Tunicates	Aplidine)	Ornithine Decarboxylase	Phase II NCT00884286 NCT02100657 Phase I NCT00788099 NCT02100657
Fungus	Plinabulin (NPI 2358)		Phase III NCT02504489 Phase II NCT00630110 Phase I NCT00322608
Alkoid	Zalypsis (PM00104)		Phase II NCT01222767
mollusk	Glembatumumab Vedotin	Microtubules	Phase II NCT02713828 NCT02487979 NCT02302339 NCT02363283 NCT01997333 NCT01156753 NCT00704158 NCT00704158 NCT00412828
Mollusk	Elisidepsin (PM02734)	Plasma Membrane Fluidity	Phase II NCT00884845
Mollusk	Brentuximab Vedotin (SGN-35)	CD30 and Microtubules	Phase II NCT01657331 NCT01461538 NCT02388490 NCT01393717 NCT01421667 NCT01805037 NCT02096042 NCT01851200 NCT02567851 NCT02423291 NCT01492088 NCT02462538
Sponges	Cytarabine (Ara-C)	DNA polymerase	approved
Sponges	Eribulin Mesylate	Microtubule	approved
Tunicates	Trabectedin (ET-743)	Minor groves of DNA	approved

[#]*Trial identifier code mentioned of each study taking each drug for specific clinical trial.*

Anti-Cancer Activities of Marine Flora

Marine flora is a major component of sea occupying large scale biomass. Marine flora can be included as micro-flora (actinobacteria, cyanobacteria, bacteria, and fungi), algae (microalgae, seaweeds) and flowering plants (mangroves and several of other halophytes) [26]. In cancer research, thousands of phytochemicals have reported to show the potential anticancer activities. National cancer institute (NCI) screened 1,14,000 extracts from 35,000 plant samples against variety of tumors till 1990 [26] and observed their utility against different types of cancer. Marine flora possesses a rich diversity of compounds in cancer prevention, mostly polyphenols, alkoids, polysaccharides *etc*.

Marine Micro-flora

Generally, bacteria, antinomycetes and fungi are the constitutive part of marine micro-flora. Secondary metabolites from marine bacteria such as sarcodictyin, discodermolide, and eleutherobin, have been reported to possess potent anticancer activities [26]. Actinomycetes, soil bacteria, are well reported in the antibiotic production, but most of them were from terrestrial origin, so there still exist a need to depict marine actinomycetes for anticancer drugs development. Recent study showed that marine actinomycetes act as a potential source of histone deacetylase inhibitors [27]. Pool of halophilic bacterial extracts from red sea brine showed apoptosis induction in cancer cell lines [28]. Marine lipopeptides extracted microbes also showed anticancer activities [3].

Algae

Blue green alga (cyanobacteria) are photosynthetic prokaryotes well known for a variety of applications related to human health. It is also a good source of food supplement due to richness in nutrients and easier digestibility. Several cyanobacteria produce a variety of secondary metabolites having potent biological activities, these belong to lipopeptides, amides, alkaloids, fatty acid and saccharides [29]. More than 50% of the marine cyanobacteria are exploited for bioactive substances related to cell death of cancer cells by inducing apoptosis [29]. In the recent studies, an acetylene-containing lipopeptide, jahanyne, marine cyanobacterium *Lyngbya* sp. inhibited the growth of human cancer cells and induced apoptosis in HeLa cells [30]. Largazole (class I-selective HDACi natural product) isolated from the marine cyanobacteria *Symploca* sp. has found antitumor effect [31].

A number of marine compounds from sea weeds have been reported to have metalloproteinases inhibitors (MMPIs) which have shown to play an importance role in cancer management [24]. Fucoxanthin, a carotenoid derived from diatoms and microalga such as marine brown seaweed showed anti-proliferation effect on cancer cells through inducing anti-angiogenesis, apoptosis and cell cycle arrest [32].

Marine Flowering and Coastal Plants

Marine halophytes and mangroves are noted for their high cancer preventive potential. Halophyte extracts have rich antioxidant activity with multiple health benefits and also source for cancer chemo-preventive agents [33]. Extracts of mangrove (*Ceriops decandra) is* known to have rich antioxidant and anticancer activities [34]. Previous study showed that *Rhizophora apiculata* (a marine mangrove plant) oil also showed to protect Benzo(a)Pyrene induced gastric cancer in Swiss albino mice [35]. Dolabrane-type of diterpenes tagalsins isolated from Chinese mangrove (genus *Ceriops*) has also recently shown an anticancer effect.

Anticancer Activities of Marine Fauna

Marine ecosystem has been recognized as one of the important sources of anti-inflammatory and anti-cancer drugs over the last decades. Apart from marine flora, marine fauna have also gained importance in the discovery of anti-cancer compounds. There have been increases in the number of preclinical and clinical anti-cancer compounds from marine fauna, which have also been considered for the clinical trials against different human malignancies (Table **2**) [36, 37]. Secondary metabolites of several marine invertebrates and vertebrates, including sponges, mollusca, tunicate, gorgonian, annelids, actinomycetes, crabs, shrimps, bryozoans, soft coral, shark and ascidians have been extensively explored for their anti-inflammatory and anti-carcinogenic activities in both *in vivo* and *in vitro* models. Secondary metabolites of these marine organisms have been found to have potent anti-proliferative activities in different cancer cell lines. These bioactive metabolites are mainly included as terpenoids, macrocyclic lactones, steroids, lipids, indole-derivatives, alkaloids, polysaccharides, chitosan, depsipeptides, pyrrols *etc*. These bioactive compounds can mediate anti-tumor activity or delay cancer progression *via* activation of apoptotic and anti-proliferative machineries. Among all the signaling pathways, aberrant activation of NF-κB and PI3K/Akt pathways play a major role in inhibiting carcinogenesis by initiation of chronic inflammatory responses. These natural marine compounds were found to inhibit cancer progression through the inhibition of these signaling pathways.

Porifera

Several poriferan species have shown the potent cytotoxic and anti-proliferative activity. A study on heteronemin (a marine sesterterpene isolated from *Hyrtios sp.* extract) has shown anti-proliferative activity in RKO-E6 and RKO cancer cell lines by the activation of the p53 signaling pathway and inhibition of the JNK pathway in RKO-E6 cells. It also inhibits NF-κB and activate apoptotic signaling cascades in chronic myelogenous leukemia cells [38]. Makaluvamine A, a potent anti-cancer compound isolated from the sponge *Zyzzya fuliginosa*, has shown an effective cytotoxic activity against several cancer cell lines including breast and colorectal cancers [39]. *Spirastrella spinispirulifera* and *Hyrtios erecta* extracted compound named as spongistatin 1, a macrocyclic lactone, was also found to induce cell death in Jurkat cells by the release of cytochrome C, Omi/HtrA2 and Smac/DIABLO from the mitochondria to the cytosol [40]. Eribulin mesylate (E7389), which is a synthetic analog of the marine natural macrolide halichondrin B, as initially isolated from the Japanese sponge *Halichondria okadai*, is a potent inhibitor of microtubule in breast cancer cells [41]. This bioactive compound is now in phase I of clinical studies [42]. Hemiasterlin, a well known tripeptides, is identified as natural products from marine sponges (*Cymbastela* sp., *Siphonochalina* sp. and *Auletta* sp.), and was found to inhibit cancer cells growth by depolymerization of microtubules and causes G2-M arrest of cell cycle [43]. Organic extracts from four sponge species (*Polymastia janeirensis, Haliclona tubifera, Mycale arcuiris and Raspailia* sp.) were found to be highly cytotoxic against three colorectal cancer lines (HT29, U373 and NCI-H460) [44].

Cnideria

The bioactive compounds isolated from soft corals (phylum: cnideria) are also known for their potent anti-cancer activity against different cancer cell lines. Sesquiterpenes, isolated from *Capnella imbricate*, showed an anti-inflammatory activity against cancer cells of different tumor origins [45, 46]. Dihydroxy-capnellene was also found to suppress the interaction of *myc* and *max* (transcription factors) [47 - 50] and thus can be used as potential therapeutic compound for the treatment of different human malignancies. Another well-known compound named as chabranol, a nor-sisquiterpene compound isolated from *Nephthea chabroli* induces reasonable cytotoxic effects against mouse lymphocytic leukemia cells [46, 51]. Crassocolides H–M (polyoxygenated cembranoids isolated from Sarcophyton crassocaule) showed potent cytotoxicity against medulloblastoma cancer cell line (Daoy cells), among which crassocolides I and M were found to be more effective [46]. Crassocolide H was also found to suppress the growth of human oral epidermoid carcinoma cells, while crassocolide L induces apoptosis in human cervical epitheloid carcinoma cells [52]. Another study on cembranolide diterpene, isolated from *Lobophytum cristagalli*, has also shown a strong inhibitory activity against different cancer cell lines [52]. *Lobophytum durum* and *Lobophytum crassum* produce durumolides A–C [53], and crassumolides A and C [54], respectively and have promising anti-inflammatory and anticancer effects. These compounds have been shown to inhibit excess generation of pro-inflammatory iNOS and COX-2 in LPS-stimulated murine macrophages [53, 54]. It has also been found that *Klyxum simplex* produces two diterpene compounds, klysimplexins H and B. These two bioactive compounds induce moderate cytotoxic effects against human cancer cell lines [55]. The cembrenoids, named flexilarins from *Sinularia flexibilis* also produce cytotoxic activity in several cancer cell lines [56]. Prostanoids (claviridic acid), another bioactive compound isolated from *Clavularia viridis* exhibited potent inhibitory activity against human gastric cancer cells (AGS) [57]. The cyclopentenone prostanoid, bromovulone III-a also showed promising anti-cancer compounds against multiple human cancer cell lines like colon, hepatocellular carcinoma and prostate cancer cell lines [58]. Steroids produced by *C. viridis* also showed cytotoxic activity against human T lymphocytes and colorectal adenocarcinoma [59]. Secondary metabolites mainly diterpenoids produced by *Clavularia koellikeri* also mediates strong cytotoxic effects against human T lymphocyte leukemia cells and human colorectal carcinoma [60]. Cespitularin C derived diterpenes was also found to exhibit anti-proliferative effects against human lung adenocarcinoma cells and mouse lymphocytic leukemia, while cespitularin E showed significant cytotoxicity against human lung adeno-carcinoma cell lines [61]. It was also found that a less active diterpene, Asterolaurin A, isolated from *Asterospicularia laurae* displayed cytotoxicity against human hepatocellular carcinoma cells [62]. Another well-known bioactive agent, punaglandins, isolated from *Telesto riisei* showed potent anti-neoplastic activity against several types of cancer cells. This bioactive compound was found to inhibit the accumulation of P53 and ubiquitin isopeptidase activity both *in vitro* and *in vivo* [63]. Apart from this, several secondary metabolites (sesquiterpenes, steroids and isishippuric acid) isolated from Gorgonians also showed significant cytotoxicity against many types of human cancers (hepatocellular, breast, leukemia, colon and lung) [64 - 67]. Moreover, few studies also reported the anti-proliferative activity of bioactive compounds (polyketides annulins A, B, and C), extracted from the marine hydroid *Garveia annulata* (order Anthoathecata), which also strongly inhibited indoleamine 2,3-dioxygenase (IDO), centrally involved in T Cell mediated immunorejection *in vitro* [68]. These annulins are found to be more effective than most tryptophan analogues recognized as IDO inhibitors. Few studies also reported the inhibitory activity of Solandelactones C, D, and G isolated from the hydroid *Solanderia secunda* (order Anthoathecata) against farnesyl protein transferase [46].

Arthropoda

A multipurpose biopolymers like chitin and chitosan act as structural constituents in the exoskeleton of crustaceans (such as crabs and shrimp) are also rich source of anti-cancer compounds. Chitosan polymers are highly biocompatible and non-toxic compounds and because of their unique chemical structure, chitosan can be easily processed into gels, membranes, beads, sponges, and scaffolds. Therefore, this type of structural feasibility makes this natural macromolecules highly efficient for cure or diagnosis of different types of cancer [69]. Few studies also revealed the potent anti-cancer activity of chitosan against different types of primary and metastatic melanoma cancer cell lines [70]. Chitosan mediates its anti-proliferative activity in these cancer cells *via* decreased adhesion of these cell lines and promotes apoptosis through the mitochondrial pathway. Wimardhani et. al. also showed the potent anti-cancer activity of low-molecular-weight chitosan (LMWC) and cisplatin combination on oral and non-cancer keratinocyte cell lines [71]. They observed that LMWC exhibited effective cytotoxic effects on Ca9-22, but not HaCaT cells. Few studies also showed that chitosan could also be used as a potential backbone for drug delivery of several anti-cancer compounds. Guo et. al. showed that chitosan-g-TPGS nanoparticles exerts a potent cytotoxic effects against MDR in breast cancer cells [72].

Bryozoa

In the last few decades, few studies also reported the anti-tumor activity of bioactive compounds isolated from extracts of a species of bryozoans (*Bugula neritina*). Bryostatins are a group of macrolide lactones, extracted from these species and are currently under investigation as anti-cancer compounds against different murine tumor cell lines [73, 74]. This bioactive compound exerts apoptosis in chronic lymphocytic leukaemia and induces cell death in different cancer cell lines synergistically in combination with other anticancer drugs [75 - 77].

Mollusca

Bioactive compounds derived from marine mollusks are also the essential source for the development of different anti-cancer drugs. It was found that Kahalalide F (cyclic depsipeptide, as isolated from marine molluca, *Elysia rufescens*) showed a promising anti-proliferative activity against topoisomerase II inhibitors treated multi-drugs resistant cancer cell lines. The anti-cancer activity of this bioactive compound was evaluated in different *in vivo* models [78, 79]. Apart from this, tyrindoleninone and 6-bromoisatin are indole compounds from marine mollusk (*Dicathais orbita*), which were also found to induce apoptosis in female reproductive cancer cell lines [80]. Few studies also reported that an alkaloid Lamellarin D (LAM-D), isolated from *Lamellaria* sp. showed potent cytotoxicity against different human malignancies. LAM-D was found to stabilize topoisomerase I with DNA covalent complexes and enhances the formation of DNA single strand breaks in cancer cells. This bioactive compound also induces nuclear apoptosis in leukemic cell lines *via* activation of intrinsic apoptotic pathway [81, 82]. Elisidepsin (PM02734, Irvalec), which is a synthetic cyclic peptide of the Kahalalide F family is currently also in clinical trials for different cancers [83, 84].

Chordata

Besides the marine invertebrates, few studies also reported the potent anti-cancer activity of secondary bioactive metabolites, extracted from marine vertebrates. Trabectidin (Yondelis), originally extracted from the Caribbean marine tunicate *Ecteinascidia turbinate*, was shown to exhibit potential cytotoxic effects against several cancer cell lines of both *in vitro* and *in vivo* models. This compound was also approved for the drug development undergoing extensive clinical trials in different parts of Europe because of its novel chemical structure and potent cytotoxic effects towards cancer cell lines [85, 86]. In 2007 and 2009, trabectedin both alone and in combination with other conventional anti-cancer drugs was authorized by the European commission for the treatment of cancer patients particularly advanced soft tissue sarcoma and ovarian cancer patients. Many clinical studies are still undergoing to evaluate the anti-tumor effects of trabectedin in different human malignancies including breast and prostate cancers. Another well-known compound known as asascididemin (ASC), an aromatic alkaloid isolated from the Mediterranean ascidian *Cystodytes dellechiajei* [87], also induces apoptosis in HL-60 and P388 leukemia cells [88]. Some research studies also showed the anticancer activity of a novel bioactive compound isolated from the ascidian [89]. Aplidine, a novel cyclic depsipeptide exerts cytotoxic effects against several cancer cell lines and currently in phase II/III clinical trials for many solid and hematologic human malignancies [90].

Marine Products as Nutraceuticals with Cancer Chemoprevention Potential

Several marine products have expanded their role in cancer prevention as well as nutraceutical values. Major part of the marine food has rich nutrients and fibers with low energy density and these possess high antioxidant potential. Marine fishes, crab, shrimps, seaweeds, *etc.* are most consumable products from marine sources, which are well known for their nutrient value. Interestingly, major of these components are also used for traditional medicine. Marine halophytes are nutritionally important source of polyunsaturated fatty acids (PUFAs) and their metabolites have antioxidant and health promoting values [33]. Marine fishes, crabs, seaweeds have extraordinary sources for vitamins and trace elements, which is required for normal human physiology. The rich antioxidant potential and a chemo-preventive potential of these foods also added an interest to health conscious people.

SCOPE, SUMMARY AND CONCLUDING REMARKS

Studies have shown the enormous anticancer potential of marine products, to act as antineoplastic agents with minimum non-target organ toxicity. Few investigations also demonstrated that the combinatorial administration of marine extracts with chemotherapeutic drugs increased survival of cancer patients, for example eribulin mesylate enhance survival of late stage breast cancer by treatment together with chemotherapeutic drugs [91]. A wide structural diversity of chemical compounds and bioactivity of marine compounds have shown a great potential to use as natural drugs or with little molecular modification for drug discovery. In this chapter, we demonstrated a number of marine natural compounds, which gained their cancer preventive potential through multiple investigations on biological models. Cancer preventive agents target cancer cells through the check point arrest at various phases of cell cycle and cell death by altering deregulated molecular machineries in cancer cells. Some extract compounds of marine biological resources exert their extraordinarily pharmacological values, through acting against numerous pathways involved in the carcinogenesis. Furthermore, the results based on clinical trials are making huge expectations on marine derived products as novel anticancer drugs. Currently FDA approved several marine compounds for their novel anticancer activity. Advantages of the natural or semi-synthesis of cancer drugs from marine resources have generated structural analogues with greater pharmacological activity and fewer side effects [92]. Thus, uses of combinatorial chemistry, bio-computation, bioinformatics applications and further high throughput screening process are highly afforded to generate potent anticancer drugs from the marine sources [92]. In addition, studies evidently showed that marine compounds have enormous ability to prevent, reverse and stop the carcinogenesis, however, there are many more studies still required to analyze the effects in the broad spectrum using mechanistic pathways evaluations for cancer.

CONFLICT OF INTEREST

The authors confirm that they have no conflict of interest to declare for this publication.

ACKNOWLEDGEMENTS

Shankar Suman acknowledges CSIR for providing senior research fellowship.

REFERENCES

[1]　Ferlay J, Soerjomataram I, Dikshit R, *et al.* Cancer incidence and mortality worldwide: sources, methods and major patterns in GLOBOCAN 2012. Int J Cancer 2015; 136(5): E359-86.
[http://dx.doi.org/10.1002/ijc.29210] [PMID: 25220842]

[2]　Lee HY, Chung KJ, Hwang IH, *et al.* Activation of p53 with ilimaquinone and ethylsmenoquinone, marine sponge metabolites, induces apoptosis and autophagy in colon cancer cells. Mar Drugs 2015; 13(1): 543-57.
[http://dx.doi.org/10.3390/md13010543] [PMID: 25603347]

[3]　Dey G, Bharti R, Dhanarajan G, *et al.* Marine lipopeptide Iturin A inhibits Akt mediated GSK3β and FoxO3a signaling and triggers apoptosis in breast cancer. Sci Rep 2015; 5: 10316.
[http://dx.doi.org/10.1038/srep10316] [PMID: 25974307]

[4]　Liu J, Ma L, Wu N, Liu G, Zheng L, Lin X. Aplysin sensitizes cancer cells to TRAIL by suppressing P38 MAPK/survivin pathway. Mar Drugs 2014; 12(9): 5072-88.
[http://dx.doi.org/10.3390/md12095072] [PMID: 25257790]

[5]　Blunt JW, Copp BR, Keyzers RA, Munro MH, Prinsep MR. Marine natural products. Nat Prod Rep 2013; 30(2): 237-323.
[http://dx.doi.org/10.1039/C2NP20112G] [PMID: 23263727]

[6]　Lopez D, Martinez-Luis S. Marine natural products with P-glycoprotein inhibitor properties. Mar Drugs 2014; 12(1): 525-46.
[http://dx.doi.org/10.3390/md12010525] [PMID: 24451193]

[7]　Huang XC, Xiao X, Zhang YK, *et al.* Lamellarin O, a pyrrole alkaloid from an Australian marine sponge, Ianthella sp., reverses BCRP mediated drug resistance in cancer cells. Mar Drugs 2014; 12(7): 3818-37.
[http://dx.doi.org/10.3390/md12073818] [PMID: 24979269]

[8]　Smith CD, Zilfou JT, Stratmann K, Patterson GM, Moore RE. Welwitindolinone analogues that reverse P-glycoprotein-mediated multiple drug resistance. Mol Pharmacol 1995; 47(2): 241-7.
[PMID: 7870031]

[9]　Aoki S, Chen ZS, Higasiyama K, Setiawan A, Akiyama S, Kobayashi M. Reversing effect of agosterol A, a spongean sterol acetate, on multidrug resistance in human carcinoma cells. Jpn J Cancer Res 2001; 92(8): 886-95.
[http://dx.doi.org/10.1111/j.1349-7006.2001.tb01177.x] [PMID: 11509122]

[10]   Chen ZS, Aoki S, Komatsu M, *et al.* Reversal of drug resistance mediated by multidrug resistance protein (MRP) 1 by dual effects of agosterol A on MRP1 function. Int J Cancer 2001; 93(1): 107-13.
       [http://dx.doi.org/10.1002/ijc.1290] [PMID: 11391629]

[11]   Shi Z, Jain S, Kim IW, *et al.* Sipholenol A, a marine-derived sipholane triterpene, potently reverses P-glycoprotein (ABCB1)-mediated multidrug resistance in cancer cells. Cancer Sci 2007; 98(9): 1373-80.
       [http://dx.doi.org/10.1111/j.1349-7006.2007.00554.x] [PMID: 17640301]

[12]   Aoki S, Cao L, Matsui K, *et al.* Kendarimide A, a novel peptide reversing P-glycoprotein-mediated multidrug resistance in tumor cells, from a marine sponge of *Haliclona* sp. Tetrahedron 2004; 60: 7053-9.
       [http://dx.doi.org/10.1016/j.tet.2003.07.020]

[13]   Khalil ZG, Huang XC, Raju R, Piggott AM, Capon RJ. Shornephine A: structure, chemical stability, and P-glycoprotein inhibitory properties of a rare diketomorpholine from an Australian marine-derived *Aspergillus* sp. J Org Chem 2014; 79(18): 8700-5.
       [http://dx.doi.org/10.1021/jo501501z] [PMID: 25158286]

[14]   Spitaler M, Utz I, Hilbe W, Hofmann J, Grunicke HH. PKC-independent modulation of multidrug resistance in cells with mutant (V185) but not wild-type (G185) P-glycoprotein by bryostatin 1. Biochem Pharmacol 1998; 56(7): 861-9.
       [http://dx.doi.org/10.1016/S0006-2952(98)00107-5] [PMID: 9774148]

[15]   Erba E, Bassano L, Di Liberti G, *et al.* Cell cycle phase perturbations and apoptosis in tumour cells induced by aplidine. Br J Cancer 2002; 86(9): 1510-7.
       [http://dx.doi.org/10.1038/sj.bjc.6600265] [PMID: 11986788]

[16]   Jiang Y, Miao ZH, Xu L, *et al.* Drug transporter-independent liver cancer cell killing by a marine steroid methyl spongoate *via* apoptosis induction. J Biol Chem 2011; 286(30): 26461-9.
       [http://dx.doi.org/10.1074/jbc.M111.232728] [PMID: 21659517]

[17]   Yang ZJ, Chee CE, Huang S, Sinicrope FA. The role of autophagy in cancer: therapeutic implications. Mol Cancer Ther 2011; 10(9): 1533-41.
       [http://dx.doi.org/10.1158/1535-7163.MCT-11-0047] [PMID: 21878654]

[18]   Akl MR, Ayoub NM, Ebrahim HY, *et al.* Araguspongine C induces autophagic death in breast cancer cells through suppression of c-Met and HER2 receptor tyrosine kinase signaling. Mar Drugs 2015; 13(1): 288-311.
       [http://dx.doi.org/10.3390/md13010288] [PMID: 25580621]

[19]   Kim MH, Joo HG. Immunostimulatory effects of fucoidan on bone marrow-derived dendritic cells. Immunol Lett 2008; 115(2): 138-43.
       [http://dx.doi.org/10.1016/j.imlet.2007.10.016] [PMID: 18077003]

[20]   Chen S, Ding R, Zhou Y, *et al.* Immunomodulatory effects of polysaccharide from marine fungus Phoma herbarum YS4108 on T cells and dendritic cells 2014.

[21]   Mistry SJ, Oh WK. New paradigms in microtubule-mediated endocrine signaling in prostate cancer. Mol Cancer Ther 2013; 12(5): 555-66.
       [http://dx.doi.org/10.1158/1535-7163.MCT-12-0871] [PMID: 23635655]

[22]   Chan A, Andreae PM, Northcote PT, Miller JH. Peloruside A inhibits microtubule dynamics in a breast cancer cell line MCF7. Invest New Drugs 2011; 29(4): 615-26.
       [http://dx.doi.org/10.1007/s10637-010-9398-2] [PMID: 20169398]

[23]   Huang X, Sun D, Pan Q, *et al.* JG6, a novel marine-derived oligosaccharide, suppresses breast cancer metastasis *via* binding to cofilin. Oncotarget 2014; 5(11): 3568-78.
       [http://dx.doi.org/10.18632/oncotarget.1959] [PMID: 25003327]

[24] Thomas NV, Kim SK. Fucoidans from marine algae as potential matrix metalloproteinase inhibitors. Adv Food Nutr Res 2014; 72: 177-93.
[http://dx.doi.org/10.1016/B978-0-12-800269-8.00010-5] [PMID: 25081083]

[25] Hideshima T, Richardson PG, Anderson KC. Mechanism of action of proteasome inhibitors and deacetylase inhibitors and the biological basis of synergy in multiple myeloma. Mol Cancer Ther 2011; 10(11): 2034-42.
[http://dx.doi.org/10.1158/1535-7163.MCT-11-0433] [PMID: 22072815]

[26] Sithranga Boopathy N, Kathiresan K. Anticancer drugs from marine flora: an overview. J Oncol 2010; 2010: 214186.

[27] Varghese TA, Jayasri MA, Suthindhiran K. Marine Actinomycetes as potential source for histone deacetylase inhibitors and epigenetic modulation. Lett Appl Microbiol 2015; 61(1): 69-76.
[http://dx.doi.org/10.1111/lam.12430] [PMID: 25880615]

[28] Sagar S, Esau L, Holtermann K, *et al.* Induction of apoptosis in cancer cell lines by the Red Sea brine pool bacterial extracts. BMC Complement Altern Med 2013; 13: 344.
[http://dx.doi.org/10.1186/1472-6882-13-344] [PMID: 24305113]

[29] Raja R, Hemaiswarya S, Ganesan V, Carvalho IS. Recent developments in therapeutic applications of Cyanobacteria. Crit Rev Microbiol 2016; 42(3): 394-405.
[PMID: 25629310]

[30] Iwasaki A, Ohno O, Sumimoto S, Ogawa H, Nguyen KA, Suenaga K. Jahanyne, an apoptosis-inducing lipopeptide from the marine *cyanobacterium Lyngbya* sp. Org Lett 2015; 17(3): 652-5.
[http://dx.doi.org/10.1021/ol5036722] [PMID: 25582897]

[31] Pilon JL, Clausen DJ, Hansen RJ, *et al.* Comparative pharmacokinetic properties and antitumor activity of the marine HDACi Largazole and Largazole peptide isostere. Cancer Chemother Pharmacol 2015; 75(4): 671-82.
[http://dx.doi.org/10.1007/s00280-015-2675-1] [PMID: 25616967]

[32] Rengarajan T, Rajendran P, Nandakumar N, Balasubramanian MP, Nishigaki I. Cancer preventive efficacy of marine carotenoid fucoxanthin: cell cycle arrest and apoptosis. Nutrients 2013; 5(12): 4978-89.
[http://dx.doi.org/10.3390/nu5124978] [PMID: 24322524]

[33] Ksouri R, Ksouri WM, Jallali I, *et al.* Medicinal halophytes: potent source of health promoting biomolecules with medical, nutraceutical and food applications. Crit Rev Biotechnol 2012; 32(4): 289-326.
[http://dx.doi.org/10.3109/07388551.2011.630647] [PMID: 22129270]

[34] Sithranga Boopathy N, Kandasamy K, Subramanian M, You-Jin J. Effect of mangrove tea extract from Ceriops decandra (Griff.) Ding Hou. on Salivary Bacterial Flora of DMBA Induced Hamster Buccal Pouch Carcinoma. Indian J Microbiol 2011; 51(3): 338-44.
[http://dx.doi.org/10.1007/s12088-011-0096-3] [PMID: 22754013]

[35] Thirunavukkarasu P, Ramanathan T, Asha S, *et al.* Gastric cancer protective effect of mangrove oil, derived from Rhizophora apiculata on benzo(a)pyrene induced cancer in albino mice. Int J Cancer Res 2015; 11: 19-31.
[http://dx.doi.org/10.3923/ijcr.2015.19.31]

[36] Newman DJ, Cragg GM. Natural products as sources of new drugs over the 30 years from 1981 to 2010. J Nat Prod 2012; 75(3): 311-35.
[http://dx.doi.org/10.1021/np200906s] [PMID: 22316239]

[37]   Mayer AM, Rodríguez AD, Berlinck RG, Fusetani N. Marine pharmacology in 20078: Marine compounds with antibacterial, anticoagulant, antifungal, anti-inflammatory, antimalarial, antiprotozoal, antituberculosis, and antiviral activities; affecting the immune and nervous system, and other miscellaneous mechanisms of action. Comp Biochem Physiol C Toxicol Pharmacol 2011; 153(2): 191-222.
[http://dx.doi.org/10.1016/j.cbpc.2010.08.008] [PMID: 20826228]

[38]   Human gene mapping 10.5. Oxford Conference (1990). Update to the 10th International Workshop on Human Gene Mapping. Cytogenet Cell Genet 1990; 55(1-4): 1-785.
[PMID: 1981499]

[39]   Wang W, Rayburn ER, Velu SE, Nadkarni DH, Murugesan S, Zhang R. *In vitro* and *in vivo* anticancer activity of novel synthetic makaluvamine analogues. Clin Cancer Res 2009; 15(10): 3511-8.
[http://dx.doi.org/10.1158/1078-0432.CCR-08-2689] [PMID: 19451594]

[40]   Schyschka L, Rudy A, Jeremias I, Barth N, Pettit GR, Vollmar AM. Spongistatin 1: a new chemosensitizing marine compound that degrades XIAP. Leukemia 2008; 22(9): 1737-45.
[http://dx.doi.org/10.1038/leu.2008.146] [PMID: 18548102]

[41]   McBride A, Butler SK. Eribulin mesylate: a novel halichondrin B analogue for the treatment of metastatic breast cancer. Am J Health Syst Pharm 2012; 69(9): 745-55.
[http://dx.doi.org/10.2146/ajhp110237] [PMID: 22517020]

[42]   Jimeno A. Eribulin: rediscovering tubulin as an anticancer target. Clin Cancer Res 2009; 15(12): 3903-5.
[http://dx.doi.org/10.1158/1078-0432.CCR-09-1023] [PMID: 19509144]

[43]   Anderson HJ, Coleman JE, Andersen RJ, Roberge M. Cytotoxic peptides hemiasterlin, hemiasterlin A and hemiasterlin B induce mitotic arrest and abnormal spindle formation. Cancer Chemother Pharmacol 1997; 39(3): 223-6.
[http://dx.doi.org/10.1007/s002800050564] [PMID: 8996524]

[44]   Monks NR, Lerner C, Henriques A, *et al.* Anticancer, antichemotactic and antimicrobial activities of marine sponges collected off the coast of Santa Catarina, southern Brazil. J Exp Mar Biol Ecol 2002; 281: 1-12.
[http://dx.doi.org/10.1016/S0022-0981(02)00380-5]

[45]   Chang CH, Wen ZH, Wang SK, Duh CY. Capnellenes from the Formosan soft coral Capnella imbricata. J Nat Prod 2008; 71(4): 619-21.
[http://dx.doi.org/10.1021/np0706116] [PMID: 18302334]

[46]   Rocha J, Peixe L, Gomes NC, Calado R. Cnidarians as a source of new marine bioactive compoundsan overview of the last decade and future steps for bioprospecting. Mar Drugs 2011; 9(10): 1860-86.
[http://dx.doi.org/10.3390/md9101860] [PMID: 22073000]

[47]   Grote D, Hänel F, Dahse HM, Seifert K. Capnellenes from the soft coral Dendronephthya rubeola. Chem Biodivers 2008; 5(9): 1683-93.
[http://dx.doi.org/10.1002/cbdv.200890157] [PMID: 18816521]

[48]   Chen QF, Liu ZP, Wang FP. Natural sesquiterpenoids as cytotoxic anticancer agents. Mini Rev Med Chem 2011; 11(13): 1153-64.
[http://dx.doi.org/10.2174/138955711797655399] [PMID: 22353224]

[49]   Peukert K, Staller P, Schneider A, Carmichael G, Hänel F, Eilers M. An alternative pathway for gene regulation by Myc. EMBO J 1997; 16(18): 5672-86.
[http://dx.doi.org/10.1093/emboj/16.18.5672] [PMID: 9312026]

[50]   Hermeking H. The MYC oncogene as a cancer drug target. Curr Cancer Drug Targets 2003; 3(3): 163-75.
[http://dx.doi.org/10.2174/1568009033481949] [PMID: 12769686]

[51]   Cheng SY, Huang KJ, Wang SK, *et al.* New terpenoids from the soft corals Sinularia capillosa and Nephthea chabroli. Org Lett 2009; 11(21): 4830-3.
[http://dx.doi.org/10.1021/ol901864d] [PMID: 19863144]

[52]   Huang HC, Chao CH, Kuo YH, Sheu JH. Crassocolides G-M, cembranoids from the Formosan soft coral Sarcophyton crassocaule. Chem Biodivers 2009; 6(8): 1232-42.
[http://dx.doi.org/10.1002/cbdv.200800142] [PMID: 19697342]

[53]   Cheng SY, Wen ZH, Chiou SF, *et al.* Durumolides A-E, anti-inflammatory and antibacterial cembranolides from the soft coral Lobophytum durum. Tetrahedron 2008; 64: 9698-04.
[http://dx.doi.org/10.1016/j.tet.2008.07.104]

[54]   Chao CH, Wen ZH, Wu YC, Yeh HC, Sheu JH. Cytotoxic and anti-inflammatory cembranoids from the soft coral Lobophytum crassum. J Nat Prod 2008; 71(11): 1819-24.
[http://dx.doi.org/10.1021/np8004584] [PMID: 18973388]

[55]   Chen BW, Wu YC, Chiang MY, *et al.* Eunicellin-based diterpenoids from the cultured soft coral Klyxum simplex. Tetrahedron 2009; 65: 7016-22.
[http://dx.doi.org/10.1016/j.tet.2009.06.047]

[56]   Lin YS, Chen CH, Liaw CC, *et al.* Cembrane diterpenoids from the Taiwanese soft coral Sinularia flexibilis. Tetrahedron 2009; 65: 9157-64.
[http://dx.doi.org/10.1016/j.tet.2009.09.031]

[57]   Lin YS, Khalil AT, Chiou SH, *et al.* Bioactive marine prostanoids from octocoral *Clavularia viridis*. Chem Biodivers 2008; 5(5): 784-92.
[http://dx.doi.org/10.1002/cbdv.200890075] [PMID: 18493965]

[58]   Chiang PC, Kung FL, Huang DM, *et al.* Induction of Fas clustering and apoptosis by coral prostanoid in human hormone-resistant prostate cancer cells. Eur J Pharmacol 2006; 542(1-3): 22-30.
[http://dx.doi.org/10.1016/j.ejphar.2006.05.030] [PMID: 16806159]

[59]   Iwashima M, Nara K, Nakamichi Y, Iguchi K. Three new chlorinated marine steroids, yonarasterols G, H and I, isolated from the okinawan soft coral, Clavularia viridis. Steroids 2001; 66(1): 25-32.
[http://dx.doi.org/10.1016/S0039-128X(00)00144-6] [PMID: 11090655]

[60]   Iwashima M, Matsumoto Y, Takahashi H, Iguchi K. New marine cembrane-type diterpenoids from the Okinawan soft coral Clavularia koellikeri. J Nat Prod 2000; 63(12): 1647-52.
[http://dx.doi.org/10.1021/np000309w] [PMID: 11141107]

[61]   Duh CY, El-Gamal AA, Wang SK, Dai CF. Novel terpenoids from the formosan soft coral Cespitularia hypotentaculata. J Nat Prod 2002; 65(10): 1429-33.
[http://dx.doi.org/10.1021/np020077w] [PMID: 12398538]

[62]   Lin YC, Abd El-Razek MH, Hwang TL, *et al.* Asterolaurins A-F, xenicane diterpenoids from the Taiwanese soft coral Asterospicularia laurae. J Nat Prod 2009; 72(11): 1911-6.
[http://dx.doi.org/10.1021/np900231e] [PMID: 19863101]

[63]   Verbitski SM, Mullally JE, Fitzpatrick FA, Ireland CM. Punaglandins, chlorinated prostaglandins, function as potent Michael receptors to inhibit ubiquitin isopeptidase activity. J Med Chem 2004; 47(8): 2062-70.
[http://dx.doi.org/10.1021/jm030448l] [PMID: 15056003]

[64]   Sheu JH, Hung KC, Wang GH, Duh CY. New cytotoxic sesquiterpenes from the gorgonian Isis hippuris. J Nat Prod 2000; 63(12): 1603-7.
[http://dx.doi.org/10.1021/np000271n] [PMID: 11141096]

[65] Gonzalez N, Barral MA, Rodriguez J, Jiménez C. New cytotoxic steroids from the gorgonian Isis hippuris. Structurer-activity studies. Tetrahedron 2001; 57: 3487-97.
[http://dx.doi.org/10.1016/S0040-4020(01)00223-X]

[66] Sheu JH, Chao CH, Wang GH, *et al.* The first A-nor-hippuristanol and two novel 4,5-secosuberosanoids from the Gorgonian Isis hippuris. Tetrahedron Lett 2004; 45: 6413-6.
[http://dx.doi.org/10.1016/j.tetlet.2004.07.001]

[67] Chao CH, Huang LF, Yang YL, *et al.* Polyoxygenated steroids from the gorgonian Isis hippuris. J Nat Prod 2005; 68(6): 880-5.
[http://dx.doi.org/10.1021/np050033y] [PMID: 15974612]

[68] Pereira A, Vottero E, Roberge M, Mauk AG, Andersen RJ. Indoleamine 2,3-dioxygenase inhibitors from the Northeastern Pacific Marine Hydroid Garveia annulata. J Nat Prod 2006; 69(10): 1496-9.
[http://dx.doi.org/10.1021/np060111x] [PMID: 17067170]

[69] Karagozlu MZ, Kim SK. Anticancer effects of chitin and chitosan derivatives. Adv Food Nutr Res 2014; 72: 215-25.
[http://dx.doi.org/10.1016/B978-0-12-800269-8.00012-9] [PMID: 25081085]

[70] Gibot L, Chabaud S, Bouhout S, Bolduc S, Auger FA, Moulin VJ. Anticancer properties of chitosan on human melanoma are cell line dependent. Int J Biol Macromol 2015; 72: 370-9.
[http://dx.doi.org/10.1016/j.ijbiomac.2014.08.033] [PMID: 25193096]

[71] Wimardhani YS, Suniarti DF, Freisleben HJ, Wanandi SI, Siregar NC, Ikeda MA. Chitosan exerts anticancer activity through induction of apoptosis and cell cycle arrest in oral cancer cells. J Oral Sci 2014; 56(2): 119-26.
[http://dx.doi.org/10.2334/josnusd.56.119] [PMID: 24930748]

[72] Guo Y, Chu M, Tan S, *et al.* Chitosan-g-TPGS nanoparticles for anticancer drug delivery and overcoming multidrug resistance. Mol Pharm 2014; 11(1): 59-70.
[http://dx.doi.org/10.1021/mp400514t] [PMID: 24229050]

[73] Hale KJ, Manaviazar S. New approaches to the total synthesis of the bryostatin antitumor macrolides. Chem Asian J 2010; 5(4): 704-54.
[http://dx.doi.org/10.1002/asia.200900634] [PMID: 20354984]

[74] Hornung RL, Pearson JW, Beckwith M, Longo DL. Preclinical evaluation of bryostatin as an anticancer agent against several murine tumor cell lines: *in vitro versus in vivo* activity. Cancer Res 1992; 52(1): 101-7.
[PMID: 1727368]

[75] Stone RM, Sariban E, Pettit GR, Kufe DW. Bryostatin 1 activates protein kinase C and induces monocytic differentiation of HL-60 cells. Blood 1988; 72(1): 208-13.
[PMID: 2455568]

[76] Hayun M, Okun E, Hayun R, *et al.* Synergistic effect of AS101 and Bryostatin-1 on myeloid leukemia cell differentiation *in vitro* and in an animal model. Leukemia 2007; 21(7): 1504-13.
[http://dx.doi.org/10.1038/sj.leu.2404746] [PMID: 17508000]

[77] Wang S, Wang Z, Dent P, Grant S. Induction of tumor necrosis factor by bryostatin 1 is involved in synergistic interactions with paclitaxel in human myeloid leukemia cells. Blood 2003; 101(9): 3648-57.
[http://dx.doi.org/10.1182/blood-2002-09-2739] [PMID: 12522001]

[78] Pardo B, Paz-Ares L, Tabernero J, *et al.* Phase I clinical and pharmacokinetic study of kahalalide F administered weekly as a 1-hour infusion to patients with advanced solid tumors. Clin Cancer Res 2008; 14(4): 1116-23.
[http://dx.doi.org/10.1158/1078-0432.CCR-07-4366] [PMID: 18281545]

[79] Provencio M, Sánchez A, Gasent J, Gómez P, Rosell R. Cancer treatments: can we find treasures at the bottom of the sea? Clin Lung Cancer 2009; 10(4): 295-300.
[http://dx.doi.org/10.3816/CLC.2009.n.041] [PMID: 19632950]

[80] Edwards V, Benkendorff K, Young F. Marine compounds selectively induce apoptosis in female reproductive cancer cells but not in primary-derived human reproductive granulosa cells. Mar Drugs 2012; 10(1): 64-83.
[http://dx.doi.org/10.3390/md10010064] [PMID: 22363221]

[81] Ballot C, Kluza J, Martoriati A, et al. Essential role of mitochondria in apoptosis of cancer cells induced by the marine alkaloid Lamellarin D. Mol Cancer Ther 2009; 8(12): 3307-17.
[http://dx.doi.org/10.1158/1535-7163.MCT-09-0639] [PMID: 19952118]

[82] Tardy C, Facompré M, Laine W, et al. Topoisomerase I-mediated DNA cleavage as a guide to the development of antitumor agents derived from the marine alkaloid lamellarin D: triester derivatives incorporating amino acid residues. Bioorg Med Chem 2004; 12(7): 1697-712.
[http://dx.doi.org/10.1016/j.bmc.2004.01.020] [PMID: 15028262]

[83] Salazar R, Cortés-Funes H, Casado E, et al. Phase I study of weekly kahalalide F as prolonged infusion in patients with advanced solid tumors. Cancer Chemother Pharmacol 2013; 72(1): 75-83.
[http://dx.doi.org/10.1007/s00280-013-2170-5] [PMID: 23645288]

[84] Serova M, de Gramont A, Bieche I, et al. Predictive factors of sensitivity to elisidepsin, a novel Kahalalide F-derived marine compound. Mar Drugs 2013; 11(3): 944-59.
[http://dx.doi.org/10.3390/md11030944] [PMID: 23519149]

[85] Carter NJ, Keam SJ. Trabectedin: a review of its use in soft tissue sarcoma and ovarian cancer. Drugs 2010; 70(3): 355-76.
[http://dx.doi.org/10.2165/11202860-000000000-00000] [PMID: 20166769]

[86] Villa FA, Gerwick L. Marine natural product drug discovery: Leads for treatment of inflammation, cancer, infections, and neurological disorders. Immunopharmacol Immunotoxicol 2010; 32(2): 228-37.
[http://dx.doi.org/10.3109/08923970903296136] [PMID: 20441539]

[87] Bonnard I, Bontemps N, Lahmy S, et al. Binding to DNA and cytotoxic evaluation of ascididemin, the major alkaloid from the Mediterranean ascidian Cystodytes dellechiajei. Anticancer Drug Des 1995; 10(4): 333-46.
[PMID: 7786398]

[88] Dassonneville L, Wattez N, Baldeyrou B, et al. Inhibition of topoisomerase II by the marine alkaloid ascididemin and induction of apoptosis in leukemia cells. Biochem Pharmacol 2000; 60(4): 527-37.
[http://dx.doi.org/10.1016/S0006-2952(00)00351-8] [PMID: 10874127]

[89] Bertanha CS, Januário AH, Alvarenga TA, et al. Quinone and hydroquinone metabolites from the ascidians of the genus Aplidium. Mar Drugs 2014; 12(6): 3608-33.
[http://dx.doi.org/10.3390/md12063608] [PMID: 24927227]

[90] Morande PE, Zanetti SR, Borge M, et al. The cytotoxic activity of Aplidin in chronic lymphocytic leukemia (CLL) is mediated by a direct effect on leukemic cells and an indirect effect on monocyte-derived cells. Invest New Drugs 2012; 30(5): 1830-40.
[http://dx.doi.org/10.1007/s10637-011-9740-3] [PMID: 21887502]

[91] Gourmelon C, Frenel JS, Campone M. Eribulin mesylate for the treatment of late-stage breast cancer. Expert Opin Pharmacother 2011; 12(18): 2883-90.
[http://dx.doi.org/10.1517/14656566.2011.637490] [PMID: 22087618]

[92] Gordaliza M. Natural products as leads to anticancer drugs. Clin Transl Oncol 2007; 9(12): 767-76.
[http://dx.doi.org/10.1007/s12094-007-0138-9] [PMID: 18158980]

CHAPTER 8

Natural Products as a Unique Source of Anti-Cancer Agents

Shinjini Singh[*]

Ex-Research Intern, Cytokine Research Laboratory, Department of Experimental Therapeutics, The University of Texas, M.D. Anderson Cancer Center, Houston, Texas 77054, USA

Abstract: Cancer is a major public health problem and the second leading cause of premature deaths worldwide, accounting for an incident rate of 2.6 million cases per year, mainly in Europe and the United States. This book chapter describes the historical aspect of cancer, its treatment modalities and history of natural compounds being used as anti-cancer agents. Role of marine natural compounds and their derivatives in cancer prevention, like, alkaloids, amine derivatives, macrolides, peptides and polypeptides are described in this chapter. Both, role of natural compounds extracted from plants and microbial sources are discussed along with their molecular targets and interactions to kill the cancer cells. Most of the medicinal compounds derived naturally are synthesized semi-synthetically for commercial purposes. They are then formulated into proper dosage increasing their costs. But for many natural compounds clinical trials are still to be carried out to validate their use in cancer therapy.

Keywords: Anti-cancer, Cancer therapy, Marine, Microbial, Nutraceuticals, Natural compounds, Plants.

INTRODUCTION

Cancer, a generic term, is defined as a disease that has a group of abnormal cells growing uncontrollably, disregarding the normal rules of cell division. Normal cells always keep getting signals, dictating the cells either to divide, differentiate into another cell or die. This proliferation could be fatal if allowed to continue or spread. Loss of growth controls leads to cancer. Loss of control can occur as a result of mutations in genes that are involved in cell-cycle control. A single event never turns a cell into cancerous one. Instead, accumulation of damage to a number of genes in a long duration of time, leads to cancer. It takes almost 25–35

---

[*] **Corresponding author Shinjini Singh**: Department of Experimental Therapeutics, The University of Texas, M.D. Anderson Cancer Center, Houston, Texas 77054, USA; Tel/Fax: 508-733-8407; E-mail: shinjini0507@gmail.com

Sahdeo Prasad & Amit Kumar Tyagi (Eds.)

years for normal cells to evolve into invasive cancerous cells. As a result of which, many years elapse between the initial events and the development of cancer [1].

Cancer is a group of more than 100 diseases, which develop in a long duration of time. It can occur in virtually any of the body's tissues, due to both, hereditary and environmental factors. So, according to current dogma, cancer is a multi-gene, multi-step disease that originates from a single abnormal cell with a mutated DNA sequence. Successive rounds of mutation and selective expansion of these cells result in tumor growth and progression, consecutively breaking through the basal membrane barrier that surrounds tissues and spreads to distant locations in the body. The phenomenon being named as metastasis.

Cancer, being a major public health problem, is the second leading cause of premature deaths worldwide, which also accounts for an incident rate of 2.6 million cases per year, mainly in Europe and United States [2, 3]. It is projected that the annual deaths due to cancer are about to increase to 13.1 million in 2030 (WHO 2012; http://www.who.int/mediacentre/factsheets/fs297/en/). In developing countries, the cancer incidence prevails by tumor types that are related to viral, genetic mutations and bacterial contamination [4]. Cancer is mostly a disease of lifestyle and is preventable, as most cancers are more prevalent in certain countries than others. For example, incidence of cancer in Unites States is much higher than in the Indian subcontinent (300 *vs.* 98 per 100.000 population) [1, 5, 6]. Hence, there is an indication that plant-based foods are more important in the diet for decreased risk of cancer.

There are evidences of cancer being as old as man, as it's found in the ancient remains of deceased humans and medical literature since the distant past. Cancer has also been noted in plants caused by virus, bacterium or fungi, and being limited by the cell wall [7]. The history of cancer goes back to the times of monarchs known as, Pharaohs, in ancient Egypt [8], Hippocrates (460-375 BC) [1] and even to the Indian system of medicine (5000 years ago), known as, Ayurveda [9, 10]. These ancient medical literatures are the evidences of the fact that physicians used to perform surgeries and also recommended natural products (especially plant products) to the patients. Even today, natural products play a major role in the treatment of cancer either directly or as derivatives (from plants, animals and microorganisms) [8].

TREATMENT MODALITIES OF CANCER

Early diagnosis, better health care facilities and developments in the therapies for cancer today, has resulted in a remarkable improvement of cancer survival [11]. But in spite of these great progresses made in the progress of cancer treatment and detection, understanding the molecular basis of cancer, there is no definitive cure by the improvements made in the therapies [12 - 14].

The classic treatments for cancer depend on the type, location, and the state of advancement of the cancer. The current paradigm for the primary treatment of cancer is by surgically removing the diagnosed solid tumor [15]. Despite the aggressive surgery measures being used for last so many decades for the treatment of cancer, the mortality rate due to cancer has not decreased to a great extent. Surgeons believe that almost every type of cancer can be treated more successfully by surgery if discovered at an early stage or localized stage [16].

Radiation therapy also remains an important component of cancer treatment. The use of X-rays as a means of cancer treatment was first appreciated after it's discovery by Wilhelm Conrad Rontgen, in 1895 [17]. It is a physical agent used to destroy cancer cells. The ionizing radiations deposit their energy in the cells of the tissue it passes through, hence killing the cancer cells by causing genetic changes in them. High-energy radiations damage the DNA of cells and block their ability to divide and proliferate further [18]. Although the radiation damages both the normal as well as abnormal cells, the goal of this therapy is to minimize the dose to the normal cells that are adjacent to the cancer cells. Moreover, the normal cells are efficient enough, in comparison to cancer cells, in repairing the damage by radiation [19]. Surgery and radiation therapy are mostly used together.

Chemotherapy is also available for the cancer treatment by some toxic compounds that target rapidly growing cancer cells, directly. Specific active proteins in cancer cell signal transduction pathways (for *eg.*, receptors and kinases) are targeted by the new chemotherapeutic drugs. These drugs are very less toxic to the normal cells. Over the years, use of many such drugs have been triumphant in the treatment of cancer. To reduce the side effects associated with these drugs, now new approaches are being studied like (a) the use of new combinations of drugs, (b) therapies targeting cancer cells using liposomal and monoclonal antibodies, (c) use of new chemo protective agents, (d) hematopoietic stem cell transplantation and also (e) the use of agents that have the potential to overcome multidrug resistance.

As stated earlier in the chapter, natural and nutritional compounds have been used for the cancer treatment and prevention throughout the history. High consumption of fruits and vegetables have been linked with the reduced risk of cancer. The cancer-inhibitory potential of nutrients and phytochemicals (from the plants) has been confirmed in considerable epidemiological and experimental studies [20]. Diets rich in fruits and vegetables have a huge role in protection from cancer. There are a lot of convincing evidences of health benefits from fruits and vegetables, whose consumption is related to a highly reduced risk of cancers of gastrointestinal tract [21 - 23].

Hormonal therapy is another very important component of the treatment strategy for cancer. Hormones are regulatory substances produced in the body and transported through tissue fluids to stimulate the growth of hormone sensitive tissues, like the breasts or prostate gland. Body's own hormones may be the cause of cancer that arises in breast or prostate tissue. So, drugs blocking the hormone production or changing the way hormones work are efficient ways of fighting cancer. Hormone therapy, is also a systemic treatment like chemotherapy, that affects the cancer cells throughout the body. In 1878, Thomas Beatson discovered that the breasts of rabbits stopped producing milk after the removal of their ovaries. Later, scientists identified that there was a considerable regression in the metastatic prostate cancer after removal of the testes. These days new classes of drugs (aromatase inhibitors, LHRH analogs) are being used to treat prostate and breast cancers.

Adjuvant therapy is also used for the cancer treatment. It is defined as the use of chemotherapy, after surgery in order to destroy the remaining cancer cells. Immunotherapy is another treatment modality, which uses the biological agents that mimic some of the natural signals in the body to control tumor growth. Interferons, interleukins, cytokines, and antigens are some of the examples of such agents, produced in the laboratory.

The development of inhibitors of the biomarkers like, epidermal growth factor receptor (EGFR), cyclooxygenase-2 (COX-2) and Ras have been very useful in the cancer chemoprevention strategies [24]. However, safety has always been a concern with the treatment of cancer patients using these pharmaceutical agents. Since the dietary compounds have potential ability to reduce the risk of cancer hence they have a very bright future as new anticancer agents [25]. These dietary compounds from the mother nature have been known to have multi-targeting properties and because cancer is a multi-genic disease these therapies are required for the treatment [1]. This chapter focuses on the main natural compounds used in cancer therapy and prevention.

HISTORY OF NATURAL COMPOUNDS AS CANCER THERAPEUTICS

Cancer therapies based on the biological substances found in nature such as botanicals, vitamins, foods or other products, are used for a number of golden ages by traditional medicine systems like Traditional Chinese medicine and Ayurvedic medicine (Traditional Indian medicine system). These traditional medicines are considered as alternative therapies in the western world today and form the basis of many modern medicines. In fact, more than 50% of the cancer therapies are derived from botanicals [26 - 28]. Functional foods are meant to be rich in nutritious content, are natural, and have bioactive chemical compounds that improves health condition, prevents diseases and have medicinal properties. These may include polyphenols, phytoestrogens, carotenoids, and fish oil. Nutraceuticals, hence are defined as the products isolated and purified from these functional foods and finally prepared in pharmaceutical forms [29, 30].

Traditional systems of medicine, like Ayurveda (science of long life), have been practiced effectively from more than 5000 years ago [9, 10]. However, the terminologies used by these systems regarding the understanding of human body and the diseases have not been adequately related to the modern medicine [31]. Ayurveda a is holistic medical system that includes in addition to the body, mind, the senses and soul, their relationships with each other and the universe in it's totality [31, 32].

There is a concept of tridosha in ayurveda which involves the regulation of three fundamental biological energies of all living systems. These three doshas are: vata, pitta and kapha, which are responsible for all physiologic and psychologic processes [31, 33, 34]. Ayurveda does include an abstract idea of cancer and diseases in general and also the treatment modalities. According to Ayurveda, diet and environmental factors, if unremedied, leads to the accumulation of the doshas in body which leads to cellular disturbances and ultimately tissue and organ system abnormalities [31, 35].

About 10,000 plant species with medicinal potential have been recognized and used in Asia as traditional medicine [36, 37]. Plants contain multiple metabolites like terpenes, polyphenols, alkaloids, polysaccharides and glycoproteins which are studies from the time of traditional medicine. These metabolites may have effect on a single molecular target or even multiple targets, depicting anticancer and anti-inflammatory responses [38]. Many botanical compounds depicting positive effects in cancer therapy have a vast study of past events, particularly in humans. Many natural compounds derived from plants, cancer types for which they are used and their status of clinical trials is given in Table **1**.

Table 1. Plant derived anti-cancer agents [178].

Compound	Cancer Type	Status
Vincristine	Breast, lung, leukemia, Lymphoma,	Phase III/IV
Vinblastine	Breast, lymphoma, germ cell, renal cancer	Phase III/IV
Paclitaxel	Ovary, breast, lung, bladder, head and neck cancer	Phase III/IV
Docetaxel	Breast and lung cancer	Phase III
Topotecan	Ovarian, lung and pediatric cancer	Phase II/III
Irinotecan	Colorectal and lung cancer	Phase II/III

Clinical trials (with their codes) of such natural compounds, carried out by National Cancer Institute (NCI) or other such institutes, are available along with their complete information. Like, clinical trial of Vinblastine sulphate at phase-III, was carried out by NCI (Code: NCI-2009-00336,CRD0000511991, COG-ACNS0332,NCT00392327) in treating young patients with newly diagnosed, previously untreated, high-risk medulloblastoma. Phase IV clinical trial for vinblastine (Code:NCI-2014-02026, EU-21011, EUDRACT-2007-004092-19, EURONET-PHL-LP1,NCT01088750) was carried out in treating young patients with stage IA or stage IIA Nodular lymphocyte-predominant Hodgkin Lymphoma. Phase II/III trial for Paclitaxel (Code:NCI-2014-00629, NCT02101788) was carried out in treating patients with recurrent or progressive low-grade ovarian cancer or peritoneal cavity cancer. Phase III trial for Irinotecan (Code:NCT00101686) was carried out in combination with three methods of administration of Fluoropyrimidine. Some of the natural compounds along with the references of their clinical trials are given in Table **2**.

Table 2. Natural compounds and their clinical trials.

Compound	Reference of Clinical Trial
Taxanes	[187]
Vinblastine	[188]
Vinorelbine	[189]
Vincristine	Phase-III;meduloblastoma;NCI-2009-00336,CDR0000511991,COG-ACNS0332,NCT00392327
Vindesine	[190]
Camptothecins	[191]
Epipodophyllotoxins	[192]
Curcumin	[193]

Terpenes, a group of natural products are widely known to play antibacterial, antineoplastic and other pharmaceutical effects [39, 40]. As the terpenes (especially triterpenes) show a broad spectrum of biological activity, there is growing interest in the treatment of cancer [41, 42]. Polyphenols (phenolic compounds) have also received a considerable attention in medicine because of it's association with disease prevention [43 - 47]. Common polyphenols like, resveratrol (grapes), quercitin (broccoli, onions, tea), curcumin (turmeric), have implications in cancer treatment and chemoprevention [26]. Saffron (stigmata of *Crocus sativus L.*) have shown promising anti-cancer effects *in vitro* and *iv vivo*, but clinical trials are yet to be done [48, 49].

Approximately 138 molecules and their associated analogues, from marine sources, have been shown to have anticancer potential. Out of these, 62% are novel compounds [50]. The marine organisms produce these compounds to protect themselves from the predators, to communicate and reproduce. More than 3000 compounds have been discovered from the marine environments that depict novel therapeutic effects. Among these, many compounds have been isolated and tested for their anti-cancer activity [51 - 53]. Several natural compounds originating or acquired from marine life are now undergoing clinical trials and that could be the potential leads in the anticancer therapy. Some of the marine anticancer molecules already reported are enumerated and explained further in this chapter. Many of these compounds are under clinical trials (Table **3**). For example, the reference for the clinical trial of Bleomycin, in the patients of inoperable cancer of cervix is: Herod J, Burton A, Buxton J, *et al.*, A randomized, prospective, phase III clinical trial of primary bleomycin, ifosfamide and cisplatin (BIP) chemotherapy followed by radiotherapy *versus* radiotherapy alone in operable cancer of the cervix. *Annals of Oncology.* 2000;11:1175-1181. Another clinical trial for Bleomycin for the patients of head and neck squamous cell carcinoma is: Pendleton KP and Grandis JR. Cisplatin-based chemotherapy options for recurrent and/or metastatic squamous cell cancer of the head and neck. Clin. Med Insights Ther. 2013;2013(5):10.4137/CMT.S10409.

Table 3. Marine derived anti-cancer agents [175].

Compound	Organism	Chemistry	Mechanism
Aaptamine	Sponge	Alkaloid	Induction of p21 and G2/M cell cycle arrest
Cortistatin A	Sponge	Alkaloid	Selective inhibition of angiogenesis
Bastadine 6	Sponge	Alkaloid	Inhibition of angiogenesis *in vitro* and *in vivo* involves apoptosis
Lamellarin D	Mollusc	Alkaloid	necrosis induction by ErbB3 protein and PI3K- Akt pathway
Bryostatin-1	Bryozoan	Macrolide	Potentiation of ara-C induced apoptosis by PKC-dependent release of TNF-α

(Table 3) contd.....

Compound	Organism	Chemistry	Mechanism
Lamellarin D	Mollusc	Alkaloid	Potent inhibition of topoisomerase I
Aeroplysinin	Sponge	Alkaloid	Induction of apoptosis on proliferating endothelial cells
HalichondrinB analogues	Sponge	Macrolide derivative	Induction of mitotic blockage and apoptosis
Dictyostatin	Sponge	Polyketide	Induction of tubulin polymerization

The ID for a phase- III NCI-supported clinical trial of Doxorubicin hydrochloride in treating young patients with mature B- Cell lymphoma are: NCI-2011-01251, NCI-2010-00129 and NCT01046825. Another phase-III clinical trial IDs supported by NCI of doxorubicin hydrochloride in treating patients with Triple-negative breast cancer are: NCI-2015-00128, NRG-BR1428 and NCT02488967. The ID for a Phase III clinical trial sponsored by NCI, in treating younger patients with previously untreated Acute Myeloid Leukemia by cytarabine and daunorubicin hydrochloride is NCT01802333.

ROLE OF MARINE NATURAL COMPOUNDS AND THEIR DERIVATIVES IN CANCER PREVENTION

Alkaloids

1. **Agelasine B:** These are the toxins acquired from marine sponges and were first reported by Nakamura *et al.* [54]. Two analogs of agelasine B were reported in 2011 and had shown to be highly toxic compounds against cancer cells [52, 55].
2. **Granulatimide and Isogranulatimide Analogs:** These are marine alkaloids isolated from the ascidian Didemnum granulatum. These perform as cell cycle G2/M checkpoint inhibitors [56 - 58]. Their 23 analogs have been evaluated by Deslandes *et al.* These analogs have shown to inhibit the growth of a number of cancer cells like, MCF-7, HS683 and PC-3 [59].
3. **Bis(indolyl)hydrazide-hydrazone Analogs:** Including sponges and tunicates, these alkaloids were acquired from the marine invertebrates [60, 61]. The cytotoxicity of a series of 14 bis(indolyl)hydrazide hydrazones, synthesized by Kumar *et al.*, was evaluated in six cancer cell lines, which included prostrate (PC-3, DU145, and LnCaP), breast (MCF and MDA-MB-231) and pancreatic cancer (PaCa2) [62].

Amine Derivatives

5-(2,4-Dimethylbenzyl) pyrrolidin-2-one (DMBPO): DMBPO was isolated from marine Streptomyces by Saurav and Kannabiran [59]. This compound showed cytotoxic activity in a dose and time dependent manner, towards HEP2 and HepG2 cell lines.

Macrolides

1. **Biselyngbyasides:** Three novel analogs of biselyngbyaside, namely, biselyngbyaside B, C and D were isolated by Morita *et al.* [63]. These were obtained from marine cyanobacterium Lyngbya, as colorless oils. Among these three compounds, only biselyngbyaside B [62] exhibited growth-inhibitory and apoptosis-inducing activity against the cell lines, HeLa S3 and HL60 [63].

2. **Bryostatins:** These belong to a family of macrolide lactones, which are acquired from invertebrate marine bryozoan *Bugula neritina* [64]. Bryostatin-1, according to NCI (National Cancer Institute), has exhibited anti-cancer activity against several cancer cell lines, like, lung, breast, ovarian, melanoma and sarcoma [65].

3. **Eribulin:** Eribulin mesylate was acquired from a marine sponge named Halichondria okadai. It is a non-taxane microtubule dynamics inhibitor and is a structurally simplified synthetic analog of the marine product, named halichondrin B [66]. Eribulin has depicted to retard the growth of malignant cells in a large number of human cancer cell lines. It has also shown significant antitumor activity in human tumor xenograft models derived from breast, colon, melanoma, ovarian and pancreatic cancers [66 - 68].

Peptides and Polypeptides

1. **Cyclic Depsipeptides:** Three new cyclodepsipeptides, namely, **neamphamides** B, C and D, were acquired from an Australian sponge named, Neamphius huxleyi. These compounds exhibited cytotoxic activities against a number of human cancer cell lines, for *eg.*, A549, HeLa, and PC3) [69 - 71].

2. **Viequeamides**, (a family of 2,2-dimethyl-3-hydroxy-7-octynoic acid (Dhoya)-containing cyclic depsipeptides), acquired from the "button" cyanobacterium (Rivularia sp.), are the compounds isolated from a predatory mollusk. They have exhibited cytotoxicity towards H460 human lung cancer cell lines [72, 73].

3. **Lagunamides A and B**, two cyclic depsipeptides, were isolated from filamentous marine cynobacterium, Lyngbya majuscule. Lagunamide A, exhibited anticancer activity against the cell lines, P388, A549, PC3 and SK-OV3 and the lagunamide B exhibited reduced cytotoxicity in the cell lines P388 and HCT8 [74 - 76].

4. **Cyclic heptapeptides:** Three new cycloheptapeptides were extracted and

reported by Chen *et al.* [77], from the fermentation extract of Acremonium persicinum SCSIO 115, which is a marine-derived fungus. They exhibited cytotoxic activity against human glioblastoma (SF-268), breast cancer (MCF-7), and lung cancer (NCI-H420) cell lines [78].

5. **Sepia Ink Oligopeptide (SIO):** It is a tripeptide acquired from Sepia esculenta. SIO has depicted to effectively inhibit the cellular proliferation of the prostate cancer cell lines like, DU-145, PC-3 and LNCaP, in a very efficient time and dose-dependent manner [78].

Phenols/Polyphenols

1. **Aeroplysinin-1:** It is a brominated tyrosine metabolite, extracted from Aplysina aerophoba, a marine sponge. It has exhibited to inhibit the growth of endothelial cells, colon carcinoma cells (HCT-116) and fibrosarcoma cells (HT-1080) in a concentration-dependent manner [79].

2. **Bromophenol Bis (2,3-dibromo-4,5-dihydroxybenzyl) ether (BDDE):** It is a marine compound acquired from two algae named Leathesia nana and Rhodomela confervoides (BDDE has shown potent cytotoxicity against the cancer cells lines like HeLa, HCT-116, HCT-8, K562; in a dose-dependent manner [80].

3. **Diphlorethohydroxycarmalol (DC):** This polyphenol compound was acquired from an edible brown alga named as Ishige Okamurae, which is found along the coast of Jeju Island, Korea. Kang *et al.*, showed that DC strongly inhibited the growth of promyelocytic leukemia cell line HL60 [81, 82].

Polysaccharides

Laminarin, a marine glucan, was obtained from Laminaria japonica (Laminariaceae) and Ecklonica kurome (Alariaceae). According to Park *et al.*, these polysaccharides inhibits the cell growth Ina dose-dependent manner and induces apoptosis in HT-29 human colon cells.

Quinones

1. **Prenylated Bromohydroquinones:** The compounds named as 7-hydroxy-cymopochromanone (PBQ1) and 7-hydroxycymopolone (PBQ2) were acquired from marine algae named as Cymopolia barbata. PBQ2 inhibited the growth of colon cancer cell lines HT29 whereas PBQ1 had no impact on any cancer cell lines investigated [83].

2. **SZ-685C:** This was acquired from the secondary metabolites of the mangrove endophytic fungus, from the South China sea [84, 85]. SZ-685C was found to be very potential in inhibiting the proliferation of the cancer cell lines derived

from breast cancer (MCF-7), human erythromyeloid leukemia (K562) and human promyelocytic leukemia (HL-60).

3. **Alterporriol L:** It was obtained from the mangrove endophytic fungus, of Alternaria species. ZJ9-6B and it's structure was found to be related to an anticancer drug, known as, epiadriamycin, which is widely used in the clinic. This derivative exhibited a dose-dependent inhibition of growth in the human breast cancer cell lines MCF-7 and MDA-MB-435 [86].

Sterols and Steroids

1. **9,11-Secosterol:** A new compound named 9,11-secosteroid (compound 103) was prepared from the ethanol extract of Formosa soft coral named, Sinularia granosa. This compound showed a potent cytotoxicity against human cancer cell lines like that of cervical epithelial carcinoma (HeLa), medulloblastoma and breast adenocarcinoma (MCF-7) [87].

2. **11-Dehydrosinulariolide:** Isolated from the soft coral Sinularia leptoclados, 11-Dehydrosinulariolide is an active compound. It has been shown to possess anti-tumor activity in the oral squamous cell carcinoma cell line named, CAL-27 [88].

3. **Diketosteroid (E)-Stigmasta-24(28)-en-3,6-dione:** This diketosteroid was obtained from the marine green alga, named as Tydemania expeditionis, which was collected in China Sea. This exhibited anti-tumor activity against the prostate cancer cell lines like, DU145, PC3 and LNCaP [89].

Terpenes

1. **10-Acetylirciformonin B:** This is a furanoterpenoid and was obtained from the Ircinia species of marine sponge. This terpenoid has shown inhibitory activity on the growth of leukemia HL60 cells [90].

2. **Phenazine Analogs:** Lavanducyanin (118), a phenazine derivative, was isolated from the fermentation broth of a Streptomyces sp. (strain: CNS284), along with two new terpenoid phenazines, namely, N-substituted brominated monoterpene phenazine (116) and N-substituted isoprenylated phenazine (117). All these three compounds have shown dose-dependent inhibitory activity on TNF-α induced NF-κB activity [91].

3. **Diterpenoid Compounds: Echinohalimane A:** It was obtained from a Formosan gorgonian Echinomuricea sp. and exhibited cytotoxic activity on human cancer cells lines like, colorectal adenocarcinoma (LoVo), colorectal adenocarcinoma (DLD-1), and acute lymphoblastic leukemia (MOLT-4) [92].

ROLE OF NATURAL COMPOUNDS FROM PLANTS IN CANCER PREVENTION

Plants supply an extensive stock of natural products that demonstrate important structural diversity and offer a large number of chemical compounds that have been used in the treatment of various illnesses since ages. It is a significant fact that about 80% of the global populace still relies on the plant derived medicines for various health problems [93]. There are strong evidences from the epidemiological and experimental studies that the plant-derived compounds, known as, phytochemicals, have shown to reduce the risk of colon cancer and have inhibited the stages of tumor development. During the last 20 years, more than 25% of drugs have been directly procured from the plants while the other 25% are these procured products, altered chemically [94]. It is also reported that there are plant-derived active ingredients in almost 74% of the most important drugs [95]. In the modern medicine, to date, more than 3000 plant species have been reported to be used in the treatment of cancer [96 - 99].

Taxanes

This was the most noteworthy plant compound found in the fight against cancer. It was discovered from the bark of a very rare plant Pacific Yew, *Taxus brevifolia* [100]. Taxol, now known as Paclitaxel, has become one of the most effective drug in the treatment of breast and cancer patients and have been approved worldwide for this use [8]. But these days it is extracted from the European Yew tree, *Taxus baccata.* New taxane analogues or formulations are being found as a result of efforts put in to improve the sustainability of taxanes and diminish the clinical resistance caused by them [101, 102]. Compounds like abraxane, CT-2103, docosahexenoic acid (DHA)-paclitaxel, are the new taxanes that have been formulated and have shown to have higher activity than paclitaxel. They even show higher activity in taxane-resistant tumors and also in the tumors that have been unresponsive to paclitaxel [103]. Unique cytotoxic activity by stabilizing microtubules rather than destabilizing them, has been exhibited by taxanes. They interfere with normal cellular functions like promoting the assembly of microtubules and preventing their depolymerization [104, 105].

Paclitaxel and docetaxel exhibit very high anti tumor activity in a number of solid tumors like, ovarian, breast, lung, head and neck, bladder, testis, and in some pediatric and hematological malignancies too [106, 107]. The taxanes are clinically very successful in treating cancer but still there are a number of significant side effects associated with them like hypersensitivity reactions and neuropathy. These side effects are controlled by the use of some prophylactic drugs.

Vinca alkaloids

These are the compounds extracted from the periwinkle plant *Catharanthus roseus*, which is also known as, *Vinca rosea*. Vincristine, vinblastine and vindesine were the first vinca alkaloids identified exhibiting the anti-tumor activity [108]. Vinca alkaloids have shown to disrupt the mitotic spindle assembly by interacting with the tubulin. They specifically bind to the β-tubulin hence blocking it's ability to polymerize with α-tubulin into the microtubules. The actively dividing cells are killed by the inhibition of mitosis in this process. The new vinca alkaloids, for example, vinorelbine and vinflunine, are weak binders as compared to vincristine and vinblastine [109]. The administration of vinca alkaloids, to the patients, is most commonly done by short IV injections while vinorelbine is the only one available for oral administration to the patients [110].

Camptothecins

Camptothecins are extracts of the Chinese tree *Camptotheca acuminata*. They are characterized by cytotoxic activity against a number of leukemias and solid tumors. Camptothecin was identified first in 1966 as an active constituent but had severe toxicity problems. So, later on two semisynthetic camptothecin analogues were formed named as, irinotecan and topotecan. These were found to be very effective in the treatment of colorectal and ovarian cancer patients [111].

Several synthetic camptothecin analogues like lurtotecan and exatecan mesylate are in the stages of clinical evaluations and are more advantageous than the classical campothecins. Camptothecins act as inhibitors of topoisomerase I [8]. Several pharmaceutical companies have performed extensive research for more effective camptothecin derivatives. Topotecan (Hycamtin®), developed by SmithKline Beecham (now Glaxo SmithKline), and Irinotecan (CPT-11; Camptosar®), originally developed by the Japanese company, Yakult Honsha, are now in clinical use [112].

Epipodophyllotoxins

These compounds are extracted from the Indian podophyllum plant known as, Podophyllum peltatum. Podophyllotoxin was used for its antihelminthic and cathartic properties by the American Indians. Two active compounds were derived after a lot of research, named as, etoposide and teniposide. These plant derived epipodophyllotoxins act as topoisomerase II inhibitors [113]. Both of them are similar in their action in inhibiting the tumor activity. The key cellular target for both the compounds is DNA topoisomerase II [114]. Etoposide have been approved for the treatment of lung cancer, ovarian and testicular cancers, lymphoma and acute myeloid leukemia. Teniposide has been approved for the tumors of central nervous system, malignant lymphoma and bladder cancer [115].

Curcumin

Turmeric has 2-9% of curcuminoids [116]. Curcumin being the most abundant curcuminoid in turmeric, provides about 75% of the total curcuminoids while demethoxycurcumin provides 10-20% and bisdemethoxycurcumin provides <5%. Curcumin extracts are also used as coloring agents in food [117].

Curcuma longa, is a member of Zingiberaceae family from which the spice turmeric is derived. Turmeric is a spice used in Indian, Southeast Asian and Middle Eastern countries [116]. The polyphenolic pigments that are fat-soluble, are known as curcuminoids that give yellow color to the turmeric. The most active curcuminoid in turmeric is curcumin. For centuries, turmeric has been used as an ayurvedic medicine for lots of ailments. There has been a growing interest in the anti-inflammatory and anti-cancer potential of turmeric in treating and preventing diseases. There are immense health benefits of curcumin, which are well known from the ancient to the modern times. It has been used as an anti-inflammatory medicine to treat a number of diseases like rheumatism, sinusitis, liver problems and lots of disorders like that of hepatic and biliary. It has been proved by a lot of studies that curcumin helps fighting infections and many cancer types.

Several *in vitro* studies have manifested curcumin as a reactive oxygen species and reactive nitrogen species, scavenger [118, 119]. Eicosanoids, potent chemical messengers are generated by the inflammatory response, in the metabolism of arachidonic acid [120]. Arachidonic acid is released by the hydrolysis of membrane phospholipids to phospholipase A2 (PLA2), This arachidonic acid, may be, metabolized by cyclooxygenases (COX) to form prostaglandins and thromboxanes, or by lipoxygenases (LOX) to form leukotrienes. Activities of PLA2, COX-2, and 5-LOX have been found to be inhibited by curcumin in the cultured cells [121].

A cell cycle comprises of a number of sequential stages that a cell passes through before being divided again. If there is a DNA damage, the cell cycle is arrested temporarily to repair the damage. On the other hand, if the damage is irreparable, the cell is signaled to die, which means to undergo apoptosis [122]. If there is a defect in cell-cycle regulation, propagation of mutations occur which eventually leads to the development of cancer. Curcumin has been depicted to induce cell-cycle arrest and apoptosis in a number of cancer cell lines grown in culture [123 - 128].

Enzymes called matrix metalloproteinases, help the cancer cells to the invade normal tissue. Curcumin has also been found to hamper the activity of these enzymes in cell culture studies [129 - 133]. The low toxicity of curcumin and its ability to induce apoptosis in a variety of cancer cell lines, makes it highly potential in cancer therapy and prevention [134].

Punica Granatum

Juice and oil extracted from *Punica granatum* (Pg) have depicted anticancer activity that interferes the tumor cell proliferation, cell cycle, invasion and angiogenesis [135]. Pg contains chemical components that possess various pharmacological and toxicological activities [136]. Oxidative stress (OS) is the process, which produces many toxic metabolites that initiate and promote different cancer types [137, 138]. Pg juice has shown to possess many antioxidants that reduce the OS to a great extent [139]. A great deal of antioxidant activity have been shown to be possessed by some flavonoids and anthocyanidins in the seed oil and juice of Pg [140]. Pg fruit extracts also show scavenging activity against hydroxyl radicals [141]. The antioxidant action of Pg is observed by its scavenging activity and also by its ability to form metal chelates [142].

Though an acute inflammation is a beneficial host response, it may also cause inflammatory bowel disease and cancer [143, 144]. Pg has depicted a great deal of potential in the inhibition of cyclooxygenase (COX) and lipooxygenase (LOX), which are key inflammatory mediators [145, 146]. A number of studies have illustrated the inhibitory effect of Pg on the production of pro-inflammatory cytokines [147, 148]. These studies illustrate that Pg plays an important role in the inhibition of the p38-mitogen-activated protein kinase (p38- MAPK) pathway and transcription factor, NF-κB (nuclear factor kappa-light-chain-enhancer of activated B cells). Activation of p38-MAPK and NF-κB are associated with increased gene expression of TNF-α, IL-1β, MCP1, iNOS, and COX-2 agents that are crucial mediators of inflammation [148].

Different types of cancers such as prostate [149, 150], breast, colon, and lungs have been shown to be inhibited by Pg [151 - 154]. Angiogenesis is meant to be a very potential target for cancer prevention because it is a very crucial process for the new blood vessels development and highly essential in supplying oxygen and nutrition to the tumorous growth and metastatic progression of cancers [155, 156].

Interestingly, Pg has shown to inhibit angiogenesis in a recent study [157]. Measurement of vascular endothelial growth factor (VEGF), IL-4, and migration inhibitory factor (MIF) in the conditioned media of estrogen sensitive (MCF-7) or estrogen resistant (MDA-MB-231) human breast cancer cells, and immortalized normal human breast epithelial cells (MCF-10A) were evaluated by Toiet *et al.*, to depict the anti-angiogenic potential of Pg. VEGF was found to be strongly decreased in MCF-10A and MCF-7, while MIF was increased in MDA-MB-231, showing significant potential for inhibitory effects of angiogenesis by Pg fractions on human umbilical vein endothelial cells (HUVEC) [157].

The proliferation of mouse mammary cancer cell line (WA4), derived from mouse MMTV-Wnt-1 mammary tumors was shown to be inhibited by the Pg extract, in a time and concentration-dependent manner. This anti-tumor activity shown by Pg extract was shown to happen through the arrest of cell cycle progression in the G0/G1 phase. Pg extract has also shown to induce apoptosis by increasing caspase-3 activity in a mouse mammary cancer cell line (WA4) [158]. Pg extracts and punic acid, an omega-5 long chain poly unsaturated fatty acid derived from Pg, have been shown to induce apoptosis in breast cancer cell line (MDA-M--231) and an estrogen sensitive cell line developed from MDA-MB-231 cells (MDA-ERalpha7) through lipid peroxidation and the PKC (Protein kinase C) signaling pathway. Disruption of the cellular mitochondrial membrane has also been exhibited by the Pg extract [159].

Myrrh

It is obtained from the dried resin of desert trees, known as, *Commiphora myrrha*. Historically, it was used along with frankincense. It was known for its anti-inflammatory and disinfectant roles and was used for treating stomach pain, indigestion, poor circulation irregular menstrual cycles and wound healing. It not only kills the cancer cells but it also kills those cells that are resistant to other anti-cancer drugs. It is not as powerful as the other plant-derived anti-cancer drugs like vincristine and vinblastine but it has an advantage that it can harm only the cancer cells and not the healthy cells [160].

A yellow oleo-gum resin existing in the stem of *Commiphora myrrha* is used worldwide for the production of myrrh, specially in China and Egypt. Myrrh comprises of volatile oil, resin, gum and bitter principles. Cytotoxic, analgesic, anti-inflammatory, anti-cancer, anti-parasitic and hypolipidemic activities have been exhibited by myrrh in a large number of studies [161 - 163]. A study used the process of hydro distillation to investigate the composition and possible anticancer activities of essential oils obtained from myrrh and frankincense. The effects of the two essential oils, independently and as a mixture, on ve tumor cell lines, MCF-7, HS-1, HepG2, HeLa and A549, were investigated using the MTT assay. The results indicated that the MCF-7 and HS-1 cell lines showed increased sensitivity to the myrrh and frankincense essential oils compared with the remaining cell lines. In addition, the anticancer effects of myrrh were markedly increased compared with those of frankincense, however, no significant synergistic effects were identified.

Green Tea

Tea is obtained from the dried leaves of the plant *Camellia sinensis*. It's effect on cancer prevention has been extensively studied worldwide. Tea is considered to be the most popular beverage consumed worldwide. The major polyphenols found in green tea are catechins, (-)-epigallocatechin-3-gallate (EGCG), (-)-epigallo-catechin (EGC), (-)-epicatechin-3- gallate (ECG), and (-)-epicatechin (EC), which display meta-5,7- dihydroxyl groups on the A ring [164] and di- or trihydroxyl groups on the B ring. These are the principal sites of antioxidant reactions. EGCG and ECG, harbor the D ring (gallate), which exhibits maximal antioxidant activity. The aforesaid characteristics of the tea polyphenols allow them to react with ROS like superoxide radical and singlet oxygen [166].

In many animal models of different human cancers, at their different sites, tea polyphenols have shown a great deal of inhibitory activity on tumorigenesis and tumor progression. There are lots of evidences that highlight the enzyme activities and signal transduction pathways of these compounds, which results in cell proliferation suppression, increased apoptosis; angiogenesis and cell invasion inhibition, and eventually inhibition of the disease development. The association between green tea polyphenol consumption and reduced cancer risk has been supported by a number of cancer prevention studies conducted on humans [167].

Tea, in the first place, was used as a medicine to cure various illnesses as it's young leaves were found to be rich in catechins. Tea originated 5000 years ago in Southwest China. In the olden times the tea leaves were rolled with milk products like cheese balls to reduce the harshness of it's taste. Drinking tea became a normal practice after the process of boiling it with water became widespread. Tea was stored as small pieces and steeped into water for consumption. It was drunk from wooden bowls in those times.

According to Traditional Chinese Medicine (TCM) tea was recommended to healthy people between 1100 BC and 200 BC. It was at the time of T'ang dynasty (618–907 AD) that tea along with it's medicinal properties tea also became an object of veneration. It became a good trade in China after that. Tea is well grown in the northern slopes of northern India, Sri Lanka, Tibet and Southern China because it's cultivation requires moist humid climates [168]. Catechins, a component of tea, has shown it's role in promoting weight loss. A study on female mice for 4 months was conducted to evaluate the anti-obesity effect of green tea (from 1% to 4% of their diets). The results showed significant suppression in food intake, body weight gain and fat tissue accumulation in the mice fed with green tea [168].

Green and black teas are major sources of bioactive flavonoids that depict antioxidant activity. The simpler catechin flavonoids in the green tea leaves are converted to complex phenolic constituents like theaflavins, in the process of fermenting the black tea. Some catechins have shown to inhibit squalene epoxidase, a key enzyme in the pathway of cholesterol biosynthesis. Theaflavin has shown to be twice as effective in blocking the activity of this enzyme. It has also been shown that catechins reduce the solubility of cholesterol in micelles, which regulate the cholesterol levels in animals fed with high levels of cholesterol or sugar in their diets [168].

Tea has also been used since ancient times due to its calming and curative effect. It is because of L-theanine, an amino acid present primarily in green and black teas, that render tranquilizing effects to the brain [169, 170]. It has been shown by a number of studies that anti-cancer activity of EGCG is due to the inhibition of mitogen-activated protein kinases (MAPK), activation of activator protein 1 (AP-1) and nuclear factor B (NF-κB), topoisomerase I and many other potential targets in the cell signaling pathways. Hence, EGCG not only provides the evidence for anti-cancer activity of the green tea but also offers new indications to discover and develop new drugs in cancer therapy [171].

ROLE OF NATURAL COMPOUNDS FROM MICROBIAL SOURCES IN CANCER PREVENTION

Over a century ago, Coley [172] observed tumors in the patients that had been accidentally infected with *Streptococcus pyogenes,* had degenerated and it is then when the tumor regression activity of bacteria was discovered and used clinically.

Bacterial infection stimulated the immune response due to this regression and it was this discovery that led to the immergence of cancer immunotherapy. Since then, a lot of research has been performed on microbes exploring their anti-neoplastic potential. As we know that microbes are easily accessible with respect to the collection, culture and fermentation, and are chemically diverse, this makes them extremely relevant source of active pharmaceutical compounds. Some of the examples of anti-cancer compounds derived from microbes and used in clinics are, anthracyclins, actinomycins, bleomycins and staurosporins [173, 174].

To stimulate immune responses, the whole bacteria can be used in their live state, attenuated or genetically modified forms, which may lead to many potential side effects. This could be avoided by using products derived from bacteria. There is a great deal of research going on regarding the use of bacterial toxins and spores and also on the use of bacteria as vectors for the gene therapy. Humans can have a lot of advantageous effects from the toxins derived by microorganisms, like destruction of the rapidly dividing cells in tumors [175]. Some anti-cancer agents derived from microbes, the cancer types in which they are used and the status of their clinical trials are enumerated in Table **4**.

Table 4. Microbe derived anti-cancer agents [178].

Compound	Cancer Type	Status
Actinomycin	Sarcoma and germ-cell tumors	Phase III/IV
Bleomycin	Cervix, head and neck cancer	Phase III/IV
Daunomycin	Leukemia	Phase III/IV
Doxorubicin	Lymphoma, breast, ovary, lung cancer	Phase III/IV
Epirubicin	Breast Cancer	Phase III/IV
Idarubicin	Breast cancer and leukemia	Phase III/IV

Antibiotics like anthracycline, bleomycin, actinomycin and mitomycin are all antitumor antibiotics and the most important chemotherapeutic agents. Daunomycin and agents related like doxorubicin and idarubicin are some of the agents from the above families that are used clinically. Anthracyclines, being the most used antitumor antibiotics in the clinics, inhibit the topoisomerase II and hence exhibiting antitumor activity [176, 177]. Screening a wide range of microorganisms led to the discovery of cyclosporine A and FK506, at Sandoz and Fujisawa Pharmaceuticals, respectively. These two drugs have been very impactful in preventing and treating graft rejection following the bone-marrow transplants and solid-organ transplants [178].

Rapamycin and its analogs have been produced by *Streptomyces hygroscopicus* and have been very potent in exhibiting immunosuppressive activity. These analogs inhibit the T-cell activation and proliferation pathways [175]. In osteosarcoma and rhabdomyosarcoma cell lines, rapamycin have shown to block the cell cycle at the middle-to-late G1 phase in T cells and B cells, as well [182].

Another example is of geldanamycin, which is a benzoquinone, inhibiting heat-shock protein HSP 90 [179]. Wortmannin, is another such drug, that is produced the fungus *Talaromyces wortmanni*. It forms a covalent complex with the active-site residue of phosphoinositide 3 kinase (PI3K), hence inhibiting the signal transduction pathways [180].

Thus it is evident that the toxins produced and evolved to kill the microorganisms have a number of physiological effects in animals or humans. In most of the cases, these compounds target the elements of signal transduction pathways and hence contemplated as novel targets for drug discovery in cancer [181].

ROLE OF NUTRACEUTICALS IN THE PREVENTION OF CANCER

Products isolated and purified from the foods, not being usually associated with the foods, and sold as medicinal forms are known as nutraceuticals. This term "nutraceutical" emerged as a result of the combination of the two terms, namely, "nutrition" and "pharmaceutical". Hence nutraceuticals are the bioactive substances in the concentrated forms, derived from different foods [182].

Resveratrol, that has antioxidant property and is obtained from the red grape, is a good example of a neutraceutical. Psyllium seed husk, a dietary fibre, reduces hypercholesterolemia, sulphoraphane from broccoli, that has chemo preventive role, and isoflavonoids from soy, that improves arterial health, are good examples of nutraceuticals. Beta-carotene, obtained from the petals of marigold is an another good example of neutraceutical. A large number of epidemiological studies have been carried out to comprehend the therapeutic effects of neutraceuticals in cancer prevention. In experimental set ups, yellow mustard oil, from the Brassica family, and mustard gum have shown to portray anti-cancer and chemo preventive roles in the advancing models of colon cancer [183].

ROLE OF FUNCTIONAL FOODS IN THE PREVENTION OF CANCER

Functional foods are the ones that demonstrate physiological benefits and hence reduce the risk of chronic diseases. They contain bioactive compounds and are consumed as a normal diet. Dietary vegetables, medicinal herbs and many plant extracts are being taken as functional foods by a large number of people these days for cancer prevention and treatment. The Indian traditional medicine, Ayurveda, has shown the path to humans in this era, for using these functional foods as agents for cancer treatment. Role of many spices used in daily cooking, especially in Indian cooking, have been proved to depict anti-cancer property in either *in vivo* or *in vitro* experimental models [184, 185]. An excellent example of an ayurvedic preparation, is Maharshi Amrit Kalash (MAK), which is composed of two components, namely, MAK-4 and MAK-5. It has illustrated anti-cancer activity on cancer cell lines *in vitro* [186].

CONCLUSION

Natural products, many of which are consumed daily with the diet, have been a principal source for the treatment of many types of cancer. These natural products provide significant protection from most of the types of cancer and many other ailments. Hence, consuming antioxidant rich fruits, vegetables and herbs provide immense health-protective effects. These anti-cancer activities of plants can be attributed to either a discrete compound or to a number of combinations of different compounds in the form of a crude extract in humans. Almost all of the natural compounds discussed in this chapter have demonstrated their anti-cancer activities *in vitro*, and most of them still have to be clinically effective in humans.

The simple methods used in the product preparation for traditional use of natural compounds in cancer treatment is relatively cheap due to the availability of plants. However, commercialization of natural compounds may deal with the problems of diminishing natural resources and adulteration in the products. Hence, most of the naturally derived medicinal compounds are synthesized semi-synthetically for commercial purposes. They are then formulated into proper dosage increasing their costs. A plethora of active agents derived from the mother nature, introduced in the cancer inventory, have changed the natural history of a number of types of human cancer.

Marine organisms produce novel pharmacological compounds with very few adverse effects. The bioactive compounds derived from marine organisms are currently of great interest to cure several types of cancers and many other ailments too. Natural product research has reported that about 80% of anticancer drugs and 45% of new drugs have been approved, derived from natural compounds. The medicinal botanicals and natural compounds are extensively investigated for their chemo preventive activities against a number of cancer types both in cell culture and animal models. Hence, before reaching the market these compounds have been tested for their chemo preventive or anti tumor activities and have already showed promise against several types of cancers.

Conclusively, we can say that compounds derived from the natural products offer us a great opportunity to evaluate new classes of anti-cancer agents and also their relevant mechanisms of action.

CONFLICT OF INTEREST

The authors confirm that they have no conflict of interest to declare for this publication.

ACKNOWLEDGEMENTS

Declared none.

REFERENCES

[1]   Singh S, Tyagi AK, Raman S, *et al.* Genome-Based Multi-targeting of Cancer: Hype or Hope?: Multi-Targeted Approach to Treatment of Cancer. Switzerland: Springerlink International Publishing 2015; pp. 19-56.
        [http://dx.doi.org/10.1007/978-3-319-12253-3_2]

[2]   Jemal A, Siegel R, Ward E, *et al.* Cancer statistics, 2008. CA Cancer J Clin 2008; 58(2): 71-96.
        [http://dx.doi.org/10.3322/CA.2007.0010] [PMID: 18287387]

[3]   Siegel R, Naishadham D, Jemal A. Cancer statistics for Hispanics/Latinos, 2012. CA Cancer J Clin 2012; 62(5): 283-98.
        [http://dx.doi.org/10.3322/caac.21153] [PMID: 22987332]

[4]   Jemal A, Siegel R, Xu J, Ward E. Cancer statistics, 2010. CA Cancer J Clin 2010; 60(5): 277-300.
        [http://dx.doi.org/10.3322/caac.20073] [PMID: 20610543]

[5]   Aggarwal BB, Kunnumakkara AB. Molecular Targets and Therapeutic Uses of Spices: Modern Uses for Ancient Medicine. World Scientific Publishing Co. Pte. Ltd. 2009.
        [http://dx.doi.org/10.1142/7150]

[6]   Goss PE, Strasser-Weippl K, Lee-Bychkovsky BL, *et al.* Challenges to effective cancer control in China, India, and Russia. Lancet Oncol 2014; 15(5): 489-538.
        [http://dx.doi.org/10.1016/S1470-2045(14)70029-4] [PMID: 24731404]

[7]   Doonan JH, Sablowski R. Walls around tumours - why plants do not develop cancer. Nat Rev Cancer 2010; 10(11): 794-802.
        [http://dx.doi.org/10.1038/nrc2942] [PMID: 20966923]

[8]   Nobili S, Lippi D, Witort E, *et al.* Natural compounds for cancer treatment and prevention. Pharmacol Res 2009; 59(6): 365-78.
        [http://dx.doi.org/10.1016/j.phrs.2009.01.017] [PMID: 19429468]

[9]   Balachandran P, Govindarajan R. Canceran ayurvedic perspective. Pharmacol Res 2005; 51(1): 19-30.
        [http://dx.doi.org/10.1016/j.phrs.2004.04.010] [PMID: 15519531]

[10]   Garodia P, Ichikawa H, Malani N, Sethi G, Aggarwal BB. From ancient medicine to modern medicine: ayurvedic concepts of health and their role in inflammation and cancer. J Soc Integr Oncol 2007; 5(1): 25-37.
        [http://dx.doi.org/10.2310/7200.2006.029] [PMID: 17309811]

[11]   Urruticoechea A, Alemany R, Balart J, Villanueva A, Viñals F, Capellá G. Recent advances in cancer therapy: an overview. Curr Pharm Des 2010; 16(1): 3-10.
        [http://dx.doi.org/10.2174/138161210789941847] [PMID: 20214614]

[12]   Hu Y, Fu L. Targeting cancer stem cells: a new therapy to cure cancer patients. Am J Cancer Res 2012; 2(3): 340-56.
        [PMID: 22679565]

[13]   Reya T, Morrison SJ, Clarke MF, Weissman IL. Stem cells, cancer, and cancer stem cells. Nature 2001; 414(6859): 105-11.
[http://dx.doi.org/10.1038/35102167] [PMID: 11689955]

[14]   Dean M, Fojo T, Bates S. Tumour stem cells and drug resistance. Nat Rev Cancer 2005; 5(4): 275-84.
[http://dx.doi.org/10.1038/nrc1590] [PMID: 15803154]

[15]   Benjamin DJ. The efficacy of surgical treatment of cancer - 20 years later. Med Hypotheses 2014; 82(4): 412-20.
[http://dx.doi.org/10.1016/j.mehy.2014.01.004] [PMID: 24480434]

[16]   Cutler DM. Are we finally winning the war on cancer? J Econ Perspect 2008; 22(4): 3-26.
[http://dx.doi.org/10.1257/jep.22.4.3] [PMID: 19768842]

[17]   Baskar R, Lee KA, Yeo R, Yeoh KW. Cancer and radiation therapy: current advances and future directions. Int J Med Sci 2012; 9(3): 193-9.
[http://dx.doi.org/10.7150/ijms.3635] [PMID: 22408567]

[18]   Jackson SP, Bartek J. The DNA-damage response in human biology and disease. Nature 2009; 461(7267): 1071-8.
[http://dx.doi.org/10.1038/nature08467] [PMID: 19847258]

[19]   Begg AC, Stewart FA, Vens C. Strategies to improve radiotherapy with targeted drugs. Nat Rev Cancer 2011; 11(4): 239-53.
[http://dx.doi.org/10.1038/nrc3007] [PMID: 21430696]

[20]   Pezzuto JM. Plant-derived anticancer agents. Biochem Pharmacol 1997; 53(2): 121-33.
[http://dx.doi.org/10.1016/S0006-2952(96)00654-5] [PMID: 9037244]

[21]   Terry P, Hu FB, Hansen H, Wolk A. Prospective study of major dietary patterns and colorectal cancer risk in women. Am J Epidemiol 2001; 154(12): 1143-9.
[http://dx.doi.org/10.1093/aje/154.12.1143] [PMID: 11744520]

[22]   Van Duyn MA, Pivonka E. Overview of the health benefits of fruit and vegetable consumption for the dietetics professional: selected literature. J Am Diet Assoc 2000; 100(12): 1511-21.
[http://dx.doi.org/10.1016/S0002-8223(00)00420-X] [PMID: 11138444]

[23]   Steinmetz KA, Potter JD. Vegetables, fruit, and cancer prevention: a review. J Am Diet Assoc 1996; 96(10): 1027-39.
[http://dx.doi.org/10.1016/S0002-8223(96)00273-8] [PMID: 8841165]

[24]   Haddad RI, Shin DM. Recent advances in head and neck cancer. N Engl J Med 2008; 359(11): 1143-54.
[http://dx.doi.org/10.1056/NEJMra0707975] [PMID: 18784104]

[25]   Gullett NP, Ruhul Amin AR, Bayraktar S, *et al.* Cancer prevention with natural compounds. Semin Oncol 2010; 37(3): 258-81.
[http://dx.doi.org/10.1053/j.seminoncol.2010.06.014] [PMID: 20709209]

[26]   Melnick SJ. Developmental therapeutics: review of biologically based CAM therapies for potential application in children with cancer: part I. J Pediatr Hematol Oncol 2006; 28(4): 221-30.
[http://dx.doi.org/10.1097/01.mph.0000212922.16427.04] [PMID: 16679919]

[27]   Müller WE G, Schröder HC, Wiens M, *et al.* Traditional and Modern Biomedical Prospecting: Part II—The Benefits. Evid Based Complement Altern Med 2004; 1(2): 133-44.
[http://dx.doi.org/10.1093/ecam/neh030]

[28]   Zakrzewski PA. Bioprospecting or biopiracy? The pharmaceutical industry's use of indigenous medicinal plants as a source of potential drug candidates. Univ Toronto Med J 2002; 79: 252-4.

[29] Biologically based practices: an overview National Center for Complementary and Alternative Medicine Available at: http://nccam.nih.gov/health/backgrounds/ biobasedprac.htm.

[30] Cencic A, Chingwaru. The role of functional foods, nutraceuticals and food supplements in intestinal health. Nutrients 2010; 2(6): 611-25.
[http://dx.doi.org/10.3390/nu2060611]

[31] Hankey A. Ayurvedic physiology and etiology: Ayurvedo Amritanaam. The doshas and their functioning in terms of contemporary biology and physical chemistry. J Altern Complement Med 2001; 7(5): 567-74.
[http://dx.doi.org/10.1089/10755530152639792] [PMID: 11719949]

[32] Subbarayappa BV. The roots of ancient medicine: an historical outline. J Biosci 2001; 26(2): 135-43.
[http://dx.doi.org/10.1007/BF02703637] [PMID: 11426049]

[33] Gerson S. Ayurvedic medicine antitoxification *versus* detoxification. Altern Complement Ther 2001; 7: 233-9.
[http://dx.doi.org/10.1089/107628001750424571]

[34] Hankey A. The scientific value of Ayurveda. J Altern Complement Med 2005; 11(2): 221-5.
[http://dx.doi.org/10.1089/acm.2005.11.221] [PMID: 15865485]

[35] Singh RH. An assessment of the ayurvedic concept of cancer and a new paradigm of anticancer treatment in Ayurveda. J Altern Complement Med 2002; 8(5): 609-14.
[http://dx.doi.org/10.1089/107555302320825129] [PMID: 12470442]

[36] Yi Y-D, Chang I-M. An overview of traditional Chinese herbal formulas and a proposal of a new code system for expressing the formula titles. Evid Based Complement Alternat Med 2004; 1(2): 125-32.
[http://dx.doi.org/10.1093/ecam/neh019] [PMID: 15480438]

[37] Seth SD, Sharma B. Medicinal plants in India. Indian J Med Res 2004; 120(1): 9-11.
[PMID: 15299226]

[38] Patocka J. Biologically active pentacyclic triterpenes and their current medicine signification. J Appl Biomed 2003; 1: 7-12.

[39] Trapp SC, Croteau RB. Genomic organization of plant terpene synthases and molecular evolutionary implications. Genetics 2001; 158(2): 811-32.
[PMID: 11404343]

[40] Ourisson G. The general role of terpenes and their global significance. Pure Appl Chem 1990; 62(7): 1401-4.
[http://dx.doi.org/10.1351/pac199062071401]

[41] Sporn MB, Suh N. Chemoprevention of cancer. Carcinogenesis 2000; 21(3): 525-30.
[http://dx.doi.org/10.1093/carcin/21.3.525] [PMID: 10688873]

[42] Patocka J. Biologically active pentacyclic triterpenes and their current medicine signification. J Appl Biomed 2003; 1: 7-12.

[43] Scalbert A, Williamson G. Dietary intake and bioavailability of polyphenols. J Nutr 2000; 130(8S) (Suppl.): 2073S-85S.
[PMID: 10917926]

[44] Youdim KA, Spencer JP, Schroeter H, Rice-Evans C. Dietary flavonoids as potential neuroprotectants. Biol Chem 2002; 383(3-4): 503-19.
[http://dx.doi.org/10.1515/BC.2002.052] [PMID: 12033439]

[45] Koo JL, Min DB. Reactive oxygen species, aging, and antioxidative nutraceuticals. Comprehensive Rev Food Sci Food Safety 2004; 3: 21-33.
[http://dx.doi.org/10.1111/j.1541-4337.2004.tb00058.x]

[46]   Finley JW. Proposed criteria for assessing the efficacy of cancer reduction by plant foods enriched in carotenoids, glucosinolates, polyphenols and selenocompounds. Ann Bot (Lond) 2005; 95(7): 1075-96.
[http://dx.doi.org/10.1093/aob/mci123] [PMID: 15784686]

[47]   Urquiaga I, Leighton F. Plant polyphenol antioxidants and oxidative stress. Biol Res 2000; 33(2): 55-64.
[http://dx.doi.org/10.4067/S0716-97602000000200004] [PMID: 15693271]

[48]   Abdullaev FI, Espinosa A. Biomedical properties of saffron and its potential use in cancer therapy and chemopreventive trials. Cancer Detect Prev 2004; 28: 426-32.

[49]   Schmidt M, Betti G, Hensel A. Saffron in phytotherapy: pharmacology and clinical uses. Wien Med Wochenschr 2007; 157(13-14): 315-9.
[http://dx.doi.org/10.1007/s10354-007-0428-4] [PMID: 17704979]

[50]   Sawadogo WR, Boly R, Cerella C, Teiten MH, Dicato M, Diederich M. A Survey of Marine Natural Compounds and Their Derivatives with Anti-cancer Activity Reported in 2012. Molecules 2015; 20(4): 7097-142.
[http://dx.doi.org/10.3390/molecules20047097] [PMID: 25903364]

[51]   Sarfaraj HM, Sheeba F, Saba A, Mohd SK. Marine natural products: A lead for anticancer. Indian J Geo-Mar Sci 2012; 41: 27-39.

[52]   Sawadogo WR, Schumacher M, Teiten MH, Cerella C, Dicato M, Diederich M. A survey of marine natural compounds and their derivatives with anti-cancer activity reported in 2011. Molecules 2013; 18(4): 3641-73.
[http://dx.doi.org/10.3390/molecules18043641] [PMID: 23529027]

[53]   Schumacher M, Kelkel M, Dicato M, Diederich M. A survey of marine natural compounds and their derivatives with anti-cancer activity reported in 2010. Molecules 2011; 16(7): 5629-46.
[http://dx.doi.org/10.3390/molecules16075629] [PMID: 21993222]

[54]   Nakamura H, Wu H, Ohizumi Y, Hirata Y. Agelasine-A, -B, -C and -D, novel bicyclic diterpenoids with a 9-methyladeninium unit possessing inhibitory effects on na,K-atpase from the okinawa sea sponge Agelas sp.1). Tetrahedron Lett 1984; 25: 2989-92.
[http://dx.doi.org/10.1016/S0040-4039(01)81345-9]

[55]   Roggen H, Charnock C, Burman R, *et al.* Antimicrobial and antineoplastic activities of agelasine analogs modified in the purine 2-position. Arch Pharm (Weinheim) 2011; 344(1): 50-5.
[http://dx.doi.org/10.1002/ardp.201000148] [PMID: 21213351]

[56]   Roberge M, Berlinck RG, Xu L, *et al.* High-throughput assay for G2 checkpoint inhibitors and identification of the structurally novel compound isogranulatimide. Cancer Res 1998; 58(24): 5701-6.
[PMID: 9865726]

[57]   Jiang X, Zhao B, Britton R, *et al.* Inhibition of Chk1 by the G2 DNA damage checkpoint inhibitor isogranulatimide. Mol Cancer Ther 2004; 3(10): 1221-7.
[PMID: 15486189]

[58]   Hugon B, Anizon F, Bailly C, *et al.* Synthesis and biological activities of isogranulatimide analogues. Bioorg Med Chem 2007; 15(17): 5965-80.
[http://dx.doi.org/10.1016/j.bmc.2007.05.073] [PMID: 17582773]

[59]   Deslandes S, Lamoral-Theys D, Frongia C, *et al.* Synthesis and biological evaluation of analogs of the marine alkaloids granulatimide and isogranulatimide. Eur J Med Chem 2012; 54: 626-36.
[http://dx.doi.org/10.1016/j.ejmech.2012.06.012] [PMID: 22809559]

[60]   Diana P, Carbone A, Barraja P, Kelter G, Fiebig HH, Cirrincione G. Synthesis and antitumor activity of 2,5-bis(3-indolyl)-furans and 3,5-bis(3-indolyl)-isoxazoles, nortopsentin analogues. Bioorg Med Chem 2010; 18(12): 4524-9.
       [http://dx.doi.org/10.1016/j.bmc.2010.04.061] [PMID: 20472437]

[61]   Carbone A, Parrino B, Barraja P, *et al.* Synthesis and antiproliferative activity of 2,5-bis(--indolyl)pyrroles, analogues of the marine alkaloid nortopsentin. Mar Drugs 2013; 11(3): 643-54.
       [http://dx.doi.org/10.3390/md11030643] [PMID: 23455514]

[62]   Kumar D, Maruthi Kumar N, Ghosh S, Shah K. Novel bis(indolyl)hydrazide-hydrazones as potent cytotoxic agents. Bioorg Med Chem Lett 2012; 22(1): 212-5.
       [http://dx.doi.org/10.1016/j.bmcl.2011.11.031] [PMID: 22123320]

[63]   Morita M, Ohno O, Teruya T, Yamori T, Inuzuka T, Suenaga K. Isolation and structures of biselyngbyasides B, C, and D from the marine *cyanobacterium Lyngbya sp.*, and the biological activities of biselyngbyasides. Tetrahedron 2012; 68: 5984-90.
       [http://dx.doi.org/10.1016/j.tet.2012.05.038]

[64]   Glickman MH, Ciechanover A. The ubiquitin-proteasome proteolytic pathway: destruction for the sake of construction. Physiol Rev 2002; 82(2): 373-428.
       [http://dx.doi.org/10.1152/physrev.00027.2001] [PMID: 11917093]

[65]   Morgan RJ Jr, Leong L, Chow W, *et al.* Phase II trial of bryostatin-1 in combination with cisplatin in patients with recurrent or persistent epithelial ovarian cancer: a California cancer consortium study. Invest New Drugs 2012; 30(2): 723-8.
       [http://dx.doi.org/10.1007/s10637-010-9557-5] [PMID: 20936324]

[66]   Pean E, Klaar S, Berglund EG, *et al.* The European medicines agency review of eribulin for the treatment of patients with locally advanced or metastatic breast cancer: summary of the scientific assessment of the committee for medicinal products for human use. Clin Cancer Res 2012; 18(17): 4491-7.
       [http://dx.doi.org/10.1158/1078-0432.CCR-11-3075] [PMID: 22829199]

[67]   Renouf DJ, Tang PA, Major P, *et al.* A phase II study of the halichondrin B analog eribulin mesylate in gemcitabine refractory advanced pancreatic cancer. Invest New Drugs 2012; 30(3): 1203-7.
       [http://dx.doi.org/10.1007/s10637-011-9673-x] [PMID: 21526355]

[68]   Mukohara T, Nagai S, Mukai H, Namiki M, Minami H. Eribulin mesylate in patients with refractory cancers: a Phase I study. Invest New Drugs 2012; 30(5): 1926-33.
       [http://dx.doi.org/10.1007/s10637-011-9741-2] [PMID: 21887501]

[69]   Yamada T, Kikuchi T, Tanaka R, Numata A. Halichoblelides B and C, potent cytotoxic macrolides from a Streptomyces species separated from a marine fish. Tetrahedron Lett 2012; 53: 2842-6.
       [http://dx.doi.org/10.1016/j.tetlet.2012.03.114]

[70]   Yamada T, Minoura K, Numata A. Halichoblelide, a potent cytotoxic macrolide from a Streptomyces species separated from a marine fish. Tetrahedron Lett 2002; 43: 1721-4.
       [http://dx.doi.org/10.1016/S0040-4039(02)00102-8]

[71]   Hood KA, West LM, Rouwé B, *et al.* Peloruside A, a novel antimitotic agent with paclitaxel-like microtubule- stabilizing activity. Cancer Res 2002; 62(12): 3356-60.
       [PMID: 12067973]

[72]   Liu J, Towle MJ, Cheng H, *et al. In vitro* and *in vivo* anticancer activities of synthetic (-)-laulimalide, a marine natural product microtubule stabilizing agent. Anticancer Res 2007; 27(3B): 1509-18.
       [PMID: 17595769]

[73] Boudreau PD, Byrum T, Liu WT, Dorrestein PC, Gerwick WH. Viequeamide A, a cytotoxic member of the kulolide superfamily of cyclic depsipeptides from a marine button cyanobacterium. J Nat Prod 2012; 75(9): 1560-70.
[http://dx.doi.org/10.1021/np300321b] [PMID: 22924493]

[74] Sorres J, Martin MT, Petek S, *et al.* Pipestelides A-C: cyclodepsipeptides from the Pacific marine sponge *Pipestela candelabra.* J Nat Prod 2012; 75(4): 759-63.
[http://dx.doi.org/10.1021/np200714m] [PMID: 22364566]

[75] Umehara M, Negishi T, Tashiro T, Nakao Y, Kimura J. Structure-related cytotoxic activity of derivatives from kulokekahilide-2, a cyclodepsipeptide in Hawaiian marine mollusk. Bioorg Med Chem Lett 2012; 22(24): 7422-5.
[http://dx.doi.org/10.1016/j.bmcl.2012.10.058] [PMID: 23127885]

[76] Tripathi A, Fang W, Leong DT, Tan LT. Biochemical studies of the lagunamides, potent cytotoxic cyclic depsipeptides from the marine cyanobacterium Lyngbya majuscula. Mar Drugs 2012; 10(5): 1126-37.
[http://dx.doi.org/10.3390/md10051126] [PMID: 22822361]

[77] Chen Z, Song Y, Chen Y, Huang H, Zhang W, Ju J. Cyclic heptapeptides, cordyheptapeptides C-E, from the marine-derived fungus Acremonium persicinum SCSIO 115 and their cytotoxic activities. J Nat Prod 2012; 75(6): 1215-9.
[http://dx.doi.org/10.1021/np300152d] [PMID: 22642609]

[78] Huang F, Yang Z, Yu D, Wang J, Li R, Ding G. Sepia ink oligopeptide induces apoptosis in prostate cancer cell lines *via* caspase-3 activation and elevation of Bax/Bcl-2 ratio. Mar Drugs 2012; 10(10): 2153-65.
[http://dx.doi.org/10.3390/md10102153] [PMID: 23170075]

[79] Martínez-Poveda B, Rodríguez-Nieto S, García-Caballero M, Medina MA, Quesada AR. The antiangiogenic compound aeroplysinin-1 induces apoptosis in endothelial cells by activating the mitochondrial pathway. Mar Drugs 2012; 10(9): 2033-46.
[http://dx.doi.org/10.3390/md10092033] [PMID: 23118719]

[80] Liu M, Zhang W, Wei J, Qiu L, Lin X. Marine bromophenol bis(2,3-dibromo-4,5-dihydroxybenzyl) ether, induces mitochondrial apoptosis in K562 cells and inhibits topoisomerase I *in vitro.* Toxicol Lett 2012; 211(2): 126-34.
[http://dx.doi.org/10.1016/j.toxlet.2012.03.771] [PMID: 22484147]

[81] Kang S-M, Kim A-D, Heo S-J, *et al.* Induction of apoptosis by diphlorethohydroxycarmalol isolated from brown alga, Ishige okamurae. J Funct Foods 2012; 4: 433-9.
[http://dx.doi.org/10.1016/j.jff.2012.02.001]

[82] Lee S-H, Choi J-I, Heo S-J, *et al.* Diphlorethohydroxycarmalol isolated from Pae (Ishige okamurae) protects high glucose-induced damage in RINm5F pancreatic β cells *via* its antioxidant effects. Food Sci Biotechnol 2012; 21: 239-46.
[http://dx.doi.org/10.1007/s10068-012-0031-3]

[83] Badal S, Gallimore W, Huang G, Tzeng TR, Delgoda R. Cytotoxic and potent CYP1 inhibitors from the marine algae Cymopolia barbata. Org Med Chem Lett 2012; 2(1): 21.
[http://dx.doi.org/10.1186/2191-2858-2-21] [PMID: 22686946]

[84] Xie G, Zhu X, Li Q, *et al.* SZ-685C, a marine anthraquinone, is a potent inducer of apoptosis with anticancer activity by suppression of the Akt/FOXO pathway. Br J Pharmacol 2010; 159(3): 689-97.
[http://dx.doi.org/10.1111/j.1476-5381.2009.00577.x] [PMID: 20128807]

[85] Zhu X, He Z, Wu J, *et al.* A marine anthraquinone SZ-685C overrides adriamycin-resistance in breast cancer cells through suppressing Akt signaling. Mar Drugs 2012; 10(4): 694-711.
[http://dx.doi.org/10.3390/md10040694] [PMID: 22690138]

[86] Huang C, Jin H, Song B, *et al.* The cytotoxicity and anticancer mechanisms of alterporriol L, a marine bianthraquinone, against MCF-7 human breast cancer cells. Appl Microbiol Biotechnol 2012; 93(2): 777-85.
[http://dx.doi.org/10.1007/s00253-011-3463-4] [PMID: 21779847]

[87] Huang CY, Su JH, Duh CY, *et al.* A new 9,11-secosterol from the soft coral Sinularia granosa. Bioorg Med Chem Lett 2012; 22(13): 4373-6.
[http://dx.doi.org/10.1016/j.bmcl.2012.05.002] [PMID: 22672798]

[88] Liu CI, Chen CC, Chen JC, *et al.* Proteomic analysis of anti-tumor effects of 11-dehydrosinulariolide on CAL-27 cells. Mar Drugs 2011; 9(7): 1254-72.
[http://dx.doi.org/10.3390/md9071254] [PMID: 21822415]

[89] Zhang JL, Tian HY, Li J, *et al.* Steroids with inhibitory activity against the prostate cancer cells and chemical diversity of marine alga Tydemania expeditionis. Fitoterapia 2012; 83(5): 973-8.
[http://dx.doi.org/10.1016/j.fitote.2012.04.019] [PMID: 22561913]

[90] Su J, Chang WB, Chen HM, *et al.* 10-acetylircifomonin B, a sponge furanoterpenoid, induces DNA damage and apoptosis in leukemia cells. Molecules 2012; 17(10): 11839-48.
[http://dx.doi.org/10.1016/j.fitote.2012.04.019] [PMID: 22561913]

[91] Kondratyuk TP, Park EJ, Yu R, *et al.* Novel marine phenazines as potential cancer chemopreventive and anti-inflammatory agents. Mar Drugs 2012; 10(2): 451-64.
[http://dx.doi.org/10.3390/md10020451] [PMID: 22412812]

[92] Chung HM, Hu LC, Yen WH, *et al.* Echinohalimane A, a bioactive halimane-type diterpenoid from a Formosan gorgonian Echinomuricea sp. (Plexauridae). Mar Drugs 2012; 10(10): 2246-53.
[http://dx.doi.org/10.3390/md10102246] [PMID: 23170081]

[93] Fulda S. Evasion of apoptosis as a cellular stress response in cancer. Int J Cell Biol 2010.
[http://dx.doi.org/10.1155/2010/370835] [PMID: 370835]

[94] Amin A, Gali-Muhtasib H, Ocker M, Schneider-Stock R. Overview of major classes of plant-derived anticancer drugs. Int J Biomed Sci 2009; 5(1): 1-11.
[PMID: 23675107]

[95] Jaganathan SK, Mandal M. Antiproliferative effects of honey and of its polyphenols: a review. J Biomed Biotech 2009.
[http://dx.doi.org/10.1155/2009/830616] [PMID: 830616]

[96] Amin AR, Kucuk O, Khuri FR, Shin DM. Perspectives for cancer prevention with natural compounds. J Clin Oncol 2009; 27(16): 2712-25.
[http://dx.doi.org/10.1200/JCO.2008.20.6235] [PMID: 19414669]

[97] Cragg GM, Newman DJ. Plants as a source of anti-cancer agents. J Ethnopharmacol 2005; 100(1-2): 72-9.
[http://dx.doi.org/10.1016/j.jep.2005.05.011] [PMID: 16009521]

[98] Elmore S. Apoptosis: a review of programmed cell death. Toxicol Pathol 2007; 35(4): 495-516.
[http://dx.doi.org/10.1080/01926230701320337] [PMID: 17562483]

[99] Koehn FE, Carter GT. The evolving role of natural products in drug discovery. Nat Rev Drug Discov 2005; 4(3): 206-20.
[http://dx.doi.org/10.1038/nrd1657] [PMID: 15729362]

[100] Wani MC, Taylor HL, Wall ME, Coggon P, McPhail AT. Plant antitumor agents. VI. The isolation and structure of taxol, a novel antileukemic and antitumor agent from Taxus brevifolia. J Am Chem Soc 1971; 93(9): 2325-7.
[http://dx.doi.org/10.1021/ja00738a045] [PMID: 5553076]

[101] ten Tije AJ, Verweij J, Loos WJ, Sparreboom A. Pharmacological effects of formulation vehicles : implications for cancer chemotherapy. Clin Pharmacokinet 2003; 42(7): 665-85.
[http://dx.doi.org/10.2165/00003088-200342070-00005] [PMID: 12844327]

[102] Hennenfent KL, Govindan R. Novel formulations of taxanes: a review. Old wine in a new bottle? Ann Oncol 2006; 17(5): 735-49.
[http://dx.doi.org/10.1093/annonc/mdj100] [PMID: 16364960]

[103] Rowinsky EK, Calvo E. Novel agents that target tublin and related elements. Semin Oncol 2006; 33(4): 421-35.
[http://dx.doi.org/10.1053/j.seminoncol.2006.04.006] [PMID: 16890797]

[104] Gueritte-Voegelein F, Guenard D, Lavelle F, Le Goff MT, *et al.* Relationships between the structure of taxol analogues and their antimitotic activity. J Med Chem 1991; 34(3): 992-8.
[http://dx.doi.org/10.1021/jm00107a017]

[105] Horwitz SB, Cohen D, Rao S, Ringel I, Shen H-J, Yang C-P. Taxol: mechanisms of action and resistance. J Natl Cancer Inst Monogr 1993; 15(15): 55-61.
[http://dx.doi.org/10.1146/annurev.med.48.1.353] [PMID: 7912530]

[106] Mekhail TM, Markman M. Paclitaxel in cancer therapy. Expert Opin Pharmacother 2002; 3(6): 755-66.
[http://dx.doi.org/10.1517/14656566.3.6.755] [PMID: 12036415]

[107] Ramaswamy B, Puhalla S. Docetaxel: a tubulin-stabilizing agent approved for the management of several solid tumors. Drugs Today (Barc) 2006; 42(4): 265-79.
[http://dx.doi.org/10.1358/dot.2006.42.4.968648] [PMID: 16703123]

[108] Fahy J. Modifications in the upper velbenamine part of the Vinca alkaloids have major implications for tubulin interacting activities. Curr Pharm Des 2001; 7(13): 1181-97.
[http://dx.doi.org/10.2174/1381612013397483] [PMID: 11472261]

[109] Ngan VK, Bellman K, Hill BT, Wilson L, Jordan MA. Mechanism of mitotic block and inhibition of cell proliferation by the semisynthetic Vinca alkaloids vinorelbine and its newer derivative vinflunine. Mol Pharmacol 2001; 60(1): 225-32.
[PMID: 11408618]

[110] Levêque D, Jehl F. Molecular pharmacokinetics of catharanthus (vinca) alkaloids. J Clin Pharmacol 2007; 47(5): 579-88.
[http://dx.doi.org/10.1177/0091270007299430] [PMID: 17442684]

[111] Malonne H, Atassi G. DNA topoisomerase targeting drugs: mechanisms of action and perspectives. Anticancer Drugs 1997; 8(9): 811-22.
[http://dx.doi.org/10.1097/00001813-199710000-00001] [PMID: 9402307]

[112] Cragg GM, Newman DJ. Plants as a source of anti-cancer agents. action and perspectives. Anticancer Drugs 1997; 8: 811-22.
[PMID: 9402307]

[113] Baldwin EL, Osheroff N. Etoposide, topoisomerase II and cancer. Curr Med Chem Anticancer Agents 2005; 5(4): 363-72.

[114] Watt PM, Hickson ID. Structure and function of type II DNA topoisomerases. Biochem J 1994; 303(Pt 3): 681-95.
[http://dx.doi.org/10.1042/bj3030681] [PMID: 7980433]

[115] Hartmann JT, Lipp H-P. Camptothecin and podophyllotoxin derivatives: inhibitors of topoisomerase I and II - mechanisms of action, pharmacokinetics and toxicity profile. Drug Saf 2006; 29(3): 209-30.
[http://dx.doi.org/10.2165/00002018-200629030-00005] [PMID: 16524321]

[116] Lechtenberg M, Quandt B, Nahrstedt A. Quantitative determination of curcuminoids in Curcuma rhizomes and rapid differentiation of Curcuma domestica Val. and Curcuma xanthorrhiza Roxb. by capillary electrophoresis. Phytochem Anal 2004; 15(3): 152-8.
[http://dx.doi.org/10.1002/pca.759] [PMID: 15202598]

[117] Cimmino A, Andolfi A, Evidente A. Phenazine as an Anticancer Agent. Microbial Phenazines: Biosynthesis, Agriculture and Health. In: Chincholkar S, Thomashow L, Eds. Berlin Heidelberg: Springer-Verlag 2013; pp. 217-44.

[118] Sreejayan , Rao MN. Nitric oxide scavenging by curcuminoids. J Pharm Pharmacol 1997; 49(1): 105-7.
[http://dx.doi.org/10.1111/j.2042-7158.1997.tb06761.x] [PMID: 9120760]

[119] Sreejayan N, Rao MN. Free radical scavenging activity of curcuminoids. Arzneimittelforschung 1996; 46(2): 169-71.
[PMID: 8720307]

[120] Steele VE, Hawk ET, Viner JL, Lubet RA. Mechanisms and applications of non-steroidal anti-inflammatory drugs in the chemoprevention of cancer. Mutat Res 2003; 523-524: 137-44.
[http://dx.doi.org/10.1016/S0027-5107(02)00329-9] [PMID: 12628511]

[121] Hong J, Bose M, Ju J, *et al.* Modulation of arachidonic acid metabolism by curcumin and related beta-diketone derivatives: effects on cytosolic phospholipase A(2), cyclooxygenases and 5-lipoxygenase. Carcinogenesis 2004; 25(9): 1671-9.
[http://dx.doi.org/10.1093/carcin/bgh165] [PMID: 15073046]

[122] Stewart ZA, Westfall MD, Pietenpol JA. Cell-cycle dysregulation and anticancer therapy. Trends Pharmacol Sci 2003; 24(3): 139-45.
[http://dx.doi.org/10.1016/S0165-6147(03)00026-9] [PMID: 12628359]

[123] Sharma RA, Gescher AJ, Steward WP. Curcumin: the story so far. Eur J Cancer 2005; 41(13): 1955-68.
[http://dx.doi.org/10.1016/j.ejca.2005.05.009] [PMID: 16081279]

[124] Duvoix A, Blasius R, Delhalle S, *et al.* Chemopreventive and therapeutic effects of curcumin. Cancer Lett 2005; 223(2): 181-90.
[http://dx.doi.org/10.1016/j.canlet.2004.09.041] [PMID: 15896452]

[125] Surh YJ, Chun KS. Cancer chemopreventive effects of curcumin. Adv Exp Med Biol 2007; 595: 149-72.
[http://dx.doi.org/10.1007/978-0-387-46401-5_5] [PMID: 17569209]

[126] Singh S, Khar A. Biological effects of curcumin and its role in cancer chemoprevention and therapy. Anticancer Agents Med Chem 2006; 6(3): 259-70.
[http://dx.doi.org/10.2174/187152006776930918] [PMID: 16712454]

[127] Kuttan G, Kumar KB, Guruvayoorappan C, Kuttan R. Antitumor, anti-invasion, and antimetastatic effects of curcumin. Adv Exp Med Biol 2007; 595: 173-84.
[http://dx.doi.org/10.1007/978-0-387-46401-5_6] [PMID: 17569210]

[128] Kunnumakkara AB, Anand P, Aggarwal BB. Curcumin inhibits proliferation, invasion, angiogenesis and metastasis of different cancers through interaction with multiple cell signaling proteins. Cancer Lett 2008; 269(2): 199-225.
[http://dx.doi.org/10.1016/j.canlet.2008.03.009] [PMID: 18479807]

[129] Banerji A, Chakrabarti J, Mitra A, Chatterjee A. Effect of curcumin on gelatinase A (MMP-2) activity in B16F10 melanoma cells. Cancer Lett 2004; 211(2): 235-42.
[http://dx.doi.org/10.1016/j.canlet.2004.02.007] [PMID: 15219947]

[130] Ohashi Y, Tsuchiya Y, Koizumi K, Sakurai H, Saiki I. Prevention of intrahepatic metastasis by curcumin in an orthotopic implantation model. Oncology 2003; 65(3): 250-8.
[http://dx.doi.org/10.1159/000074478] [PMID: 14657599]

[131] Menon LG, Kuttan R, Kuttan G. Anti-metastatic activity of curcumin and catechin. Cancer Lett 1999; 141(1-2): 159-65.
[http://dx.doi.org/10.1016/S0304-3835(99)00098-1] [PMID: 10454257]

[132] Mitra A, Chakrabarti J, Banerji A, Chatterjee A, Das BR. Curcumin, a potential inhibitor of MMP-2 in human laryngeal squamous carcinoma cells HEp2. J Environ Pathol Toxicol Oncol 2006; 25(4): 679-90.
[http://dx.doi.org/10.1615/JEnvironPatholToxicolOncol.v25.i4.70] [PMID: 17341208]

[133] Hong JH, Ahn KS, Bae E, Jeon SS, Choi HY. The effects of curcumin on the invasiveness of prostate cancer *in vitro* and *in vivo*. Prostate Cancer Prostatic Dis 2006; 9(2): 147-52.
[http://dx.doi.org/10.1038/sj.pcan.4500856] [PMID: 16389264]

[134] Karunagaran D, Rashmi R, Kumar TR. Induction of apoptosis by curcumin and its implications for cancer therapy. Curr Cancer Drug Targets 2005; 5(2): 117-29.
[http://dx.doi.org/10.2174/1568009053202081] [PMID: 15810876]

[135] Lansky EP, Newman RA. Punica granatum (pomegranate) and its potential for prevention and treatment of inflammation and cancer. J Ethnopharmacol 2007; 109(2): 177-206.
[http://dx.doi.org/10.1016/j.jep.2006.09.006] [PMID: 17157465]

[136] Seeram NP, Schulman RN, Heber D. Pomegranates: Ancient Roots to Modern Medicine. Boca Raton: Taylor and Francis Group 2006; pp. 5-8.

[137] Ohshima H, Tazawa H, Sylla BS, Sawa T. Prevention of human cancer by modulation of chronic inflammatory processes. Mutat Res 2005; 591(1-2): 110-22.
[http://dx.doi.org/10.1016/j.mrfmmm.2005.03.030] [PMID: 16083916]

[138] Garcea G, Dennison AR, Steward WP, Berry DP. Role of inflammation in pancreatic carcinogenesis and the implications for future therapy. Pancreatology 2005; 5(6): 514-29.
[http://dx.doi.org/10.1159/000087493] [PMID: 16110250]

[139] Noda Y, Kaneyuki T, Mori A, Packer L. Antioxidant activities of pomegranate fruit extract and its anthocyanidins: delphinidin, cyanidin, and pelargonidin. J Agric Food Chem 2002; 50(1): 166-71.
[http://dx.doi.org/10.1021/jf0108765] [PMID: 11754562]

[140] Mori-Okamoto J, Otawara-Hamamoto Y, Yamato H, Yoshimura H. Pomegranate extract improves a depressive state and bone properties in menopausal syndrome model ovariectomized mice. J Ethnopharmacol 2004; 92(1): 93-101.
[http://dx.doi.org/10.1016/j.jep.2004.02.006] [PMID: 15099854]

[141] Guo S, Deng Q, Xiao J, Xie B, Sun Z. Evaluation of antioxidant activity and preventing DNA damage effect of pomegranate extracts by chemiluminescence method. J Agric Food Chem 2007; 55(8): 3134-40.
[http://dx.doi.org/10.1021/jf063443g] [PMID: 17381116]

[142] Kulkarni AP, Mahal HS, Kapoor S, Aradhya SM. *In vitro* studies on the binding, antioxidant, and cytotoxic actions of punicalagin. J Agric Food Chem 2007; 55(4): 1491-500.
[http://dx.doi.org/10.1021/jf0626720] [PMID: 17243704]

[143] Balkwill F, Charles KA, Mantovani A. Smoldering and polarized inflammation in the initiation and promotion of malignant disease. Cancer Cell 2005; 7(3): 211-7.
[http://dx.doi.org/10.1016/j.ccr.2005.02.013] [PMID: 15766659]

[144] Simmons DL, Buckley CD. Some new and not so new, anti-inflammatory targets. Curr Opin Pharmacol 2005; 5: 394-7.
[http://dx.doi.org/10.1016/j.coph.2005.05.001]

[145] Schubert SY, Lansky EP, Neeman I. Antioxidant and eicosanoid enzyme inhibition properties of pomegranate seed oil and fermented juice flavonoids. J Ethnopharmacol 1999; 66(1): 11-7.
[http://dx.doi.org/10.1016/S0378-8741(98)00222-0] [PMID: 10432202]

[146] Rahimi HR, Arasoo M, Shiri M. *Punica granatum* is more effective to prevent gastric disorders induced by Helicobacter pylori or any other stimulator in humans. Asian J Plant Sci 2011; 10: 380-2.
[http://dx.doi.org/10.3923/ajps.2011.380.382]

[147] Mix KS, Mengshol JA, Benbow U, Vincenti MP, Sporn MB, Brinckerhoff CE. A synthetic triterpenoid selectively inhibits the induction of matrix metalloproteinases 1 and 13 by inflammatory cytokines. Arthritis Rheum 2001; 44(5): 1096-104.
[http://dx.doi.org/10.1002/1529-0131(200105)44:5<1096::AID-ANR190>3.0.CO;2-6]   [PMID: 11352241]

[148] Hayden MS, Ghosh S. Signaling to NF-kappaB. Genes Dev 2004; 18(18): 2195-224.
[http://dx.doi.org/10.1101/gad.1228704] [PMID: 15371334]

[149] Koyama S, Cobb LJ, Mehta HH, *et al.* Pomegranate extract induces apoptosis in human prostate cancer cells by modulation of the IGF-IGFBP axis. Growth Horm IGF Res 2010; 20(1): 55-62.
[http://dx.doi.org/10.1016/j.ghir.2009.09.003] [PMID: 19853487]

[150] Rettig MB, Heber D, An J, *et al.* Pomegranate extract inhibits androgen-independent prostate cancer growth through a nuclear factor-kappaB-dependent mechanism. Mol Cancer Ther 2008; 7(9): 2662-71.
[http://dx.doi.org/10.1158/1535-7163.MCT-08-0136] [PMID: 18790748]

[151] Sturgeon SR, Ronnenberg AG. Pomegranate and breast cancer: possible mechanisms of prevention. Nutr Rev 2010; 68(2): 122-8.
[http://dx.doi.org/10.1111/j.1753-4887.2009.00268.x] [PMID: 20137057]

[152] Kasimsetty SG, Bialonska D, Reddy MK, Ma G, Khan SI, Ferreira D. Colon cancer chemopreventive activities of pomegranate ellagitannins and urolithins. J Agric Food Chem 2010; 58(4): 2180-7.
[http://dx.doi.org/10.1021/jf903762h] [PMID: 20112993]

[153] Khan SA. The role of pomegranate (Punica granatum L.) in colon cancer. Pak J Pharm Sci 2009; 22(3): 346-8.
[PMID: 19553187]

[154] Khan N, Afaq F, Kweon MH, Kim K, Mukhtar H. Oral consumption of pomegranate fruit extract inhibits growth and progression of primary lung tumors in mice. Cancer Res 2007; 67(7): 3475-82.
[http://dx.doi.org/10.1158/0008-5472.CAN-06-3941] [PMID: 17389758]

[155] Pfeffer U, Ferrari N, Morini M, Benelli R, Noonan DM, Albini A. Antiangiogenic activity of chemopreventive drugs. Int J Biol Markers 2003; 18(1): 70-4.
[http://dx.doi.org/10.5301/JBM.2008.5042] [PMID: 12699068]

[156] Scappaticci FA. The therapeutic potential of novel antiangiogenic therapies. Expert Opin Investig Drugs 2003; 12(6): 923-32.
[http://dx.doi.org/10.1517/13543784.12.6.923] [PMID: 12783597]

[157] Toi M, Bando H, Ramachandran C, *et al.* Preliminary studies on the anti-angiogenic potential of pomegranate fractions *in vitro* and *in vivo*. Angiogenesis 2003; 6(2): 121-8.
[http://dx.doi.org/10.1023/B:AGEN.0000011802.81320.e4] [PMID: 14739618]

[158] Dai Z, Nair V, Khan M, Ciolino HP. Pomegranate extract inhibits the proliferation and viability of MMTV-Wnt-1 mouse mammary cancer stem cells *in vitro*. Oncol Rep 2010; 24(4): 1087-91.
[PMID: 20811693]

[159] Grossmann ME, Mizuno NK, Schuster T, Cleary MP. Punicic acid is an omega-5 fatty acid capable of inhibiting breast cancer proliferation. Int J Oncol 2010; 36(2): 421-6.
[PMID: 20043077]

[160] Kinghorn AD, Farnsworth NR, Soejarto DD, *et al.* Novel strategies for the discovery of plant-derived anticancer agents. Pharm Biol 2003; 41: 53-67.
[http://dx.doi.org/10.1080/1388020039051744]

[161] Su S, Wang T, Duan JA, *et al.* Anti-inflammatory and analgesic activity of different extracts of *Commiphora myrrha.* J Ethnopharmacol 2011; 134(2): 251-8.
[http://dx.doi.org/10.1016/j.jep.2010.12.003] [PMID: 21167270]

[162] Wu XS, Xie T, Lin J, *et al.* An investigation of the ability of elemene to pass through the blood-brain barrier and its effect on brain carcinomas. J Pharm Pharmacol 2009; 61(12): 1653-6.
[http://dx.doi.org/10.1211/jpp.61.12.0010] [PMID: 19958588]

[163] Shoemaker M, Hamilton B, Dairkee SH, Cohen I, Campbell MJ. *In vitro* anticancer activity of twelve Chinese medicinal herbs. Phytother Res 2005; 19(7): 649-51.
[http://dx.doi.org/10.1002/ptr.1702] [PMID: 16161030]

[164] Balentine DA, Wiseman SA, Bouwens LC. The chemistry of tea flavonoids. Crit Rev Food Sci Nutr 1997; 37(8): 693-704.
[http://dx.doi.org/10.1080/10408399709527797] [PMID: 9447270]

[165] Sang S, Tian S, Meng X, *et al.* Theadibenzotropolone A, a new type pigment from enzymatic oxidation of (-)-epicatechin and (-)-epigallocatechin gallate and characterized from black tea using LC/MS/MS. Tetrahedron Lett 2002; 43(40): 7129-33.
[http://dx.doi.org/10.1016/S0040-4039(02)01707-0]

[166] Lipinski CA, Lombardo F, Dominy BW, Feeney PJ. Experimental and computational approaches to estimate solubility and permeability in drug discovery and development settings. Adv Drug Deliv Rev 2001; 46(1-3): 3-26.
[http://dx.doi.org/10.1016/S0169-409X(00)00129-0] [PMID: 11259830]

[167] Mak JC. Potential role of green tea catechins in various disease therapies: progress and promise. Clin Exp Pharmacol Physiol 2012; 39(3): 265-73.
[http://dx.doi.org/10.1111/j.1440-1681.2012.05673.x] [PMID: 22229384]

[168] Cooper R, Morré DJ, Morré DM. Medicinal benefits of green tea: Part I. Review of noncancer health benefits. J Altern Complement Med 2005; 11(3): 521-8.
[http://dx.doi.org/10.1089/acm.2005.11.521] [PMID: 15992239]

[169] Talbott SM. The Cortisol Connection. New York: Hunter House 2002.

[170] Huber LG. Green tea catechins and L-theanine in integrative. cancer care: A review of the research. Altern Complement Ther 2003; 9: 294-8.
[http://dx.doi.org/10.1089/107628003322658557]

[171] Chen L, Zhang HY. Cancer preventive mechanisms of the green tea polyphenol (-)-epigallocatechi--3-gallate. Molecules 2007; 12(5): 946-57.
[http://dx.doi.org/10.3390/12050946] [PMID: 17873830]

[172] Hoption Cann SA, van Netten JP, van Netten C. Dr William Coley and tumour regression: a place in history or in the future. Postgrad Med J 2003; 79(938): 672-80.
[PMID: 14707241]

[173] Cragg GM, Newman DJ. Natural products: a continuing source of novel drug leads. Biochim Biophys Acta 2013; 1830(6): 3670-95.
[http://dx.doi.org/10.1016/j.bbagen.2013.02.008] [PMID: 23428572]

[174] Heinrich MB, Gibbons S, Williamson EM. Fundamentals of Pharmacognosy and Phytotherapy. 1st ed., Edinburgh, UK: Churchill Livingstone 2004.

[175] Bhanot A, Sharma R, Noolvi MN. Natural sources as potential anti-cancer agents: A review. Int Phytomed 2011; 3: 9-26.

[176] Gupta SC, Patchva S, Koh W, Aggarwal BB. Discovery of curcumin, a component of golden spice, and its miraculous biological activities. Clin Exp Pharmacol Physiol 2012; 39(3): 283-99.
[http://dx.doi.org/10.1111/j.1440-1681.2011.05648.x] [PMID: 22118895]

[177] Anand P, Sundaram C, Jhurani S, Kunnumakkara AB, Aggarwal BB. Curcumin and cancer: an old-age disease with an age-old solution. Cancer Lett 2008; 267(1): 133-64.
[http://dx.doi.org/10.1016/j.canlet.2008.03.025] [PMID: 18462866]

[178] da Rocha AB, Lopes RM, Schwartsmann G. Natural products in anticancer therapy. Curr Opin Pharmacol 2001; 1(4): 364-9.
[http://dx.doi.org/10.1016/S1471-4892(01)00063-7] [PMID: 11710734]

[179] Schulte TW, Neckers LM. The benzoquinone ansamycin 17-allylamino-17-demethoxygeldanamycin binds to HSP90 and shares important biologic activities with geldanamycin. Cancer Chemother Pharmacol 1998; 42(4): 273-9.
[http://dx.doi.org/10.1007/s002800050817] [PMID: 9744771]

[180] Cardenas ME, Sanfridson A, Cutler NS, Heitman J. Signal-transduction cascades as targets for therapeutic intervention by natural products. Trends Biotechnol 1998; 16(10): 427-33.
[http://dx.doi.org/10.1016/S0167-7799(98)01239-6] [PMID: 9807840]

[181] Adjei AA. Signal transduction pathway targets for anticancer drug discovery. Curr Pharm Des 2000; 6(4): 361-78.
[http://dx.doi.org/10.2174/1381612003400821] [PMID: 10788587]

[182] Foster BC, Arnason JT, Briggs CJ. Natural health products and drug disposition. Annu Rev Pharmacol Toxicol 2005; 45: 203-26.
[http://dx.doi.org/10.1146/annurev.pharmtox.45.120403.095950] [PMID: 15822175]

[183] Eskin NA, Raju J, Bird RP. Novel mucilage fraction of *Sinapis alba* L. (mustard) reduces azoxymethane-induced colonic aberrant crypt foci formation in F344 and Zucker obese rats. Phytomedicine 2007; 14(7-8): 479-85.
[http://dx.doi.org/10.1016/j.phymed.2006.09.016] [PMID: 17188481]

[184] Tripathi YB, Tripathi P, Arjmandi BH. Nutraceuticals and cancer management. Front Biosci 2005; 10: 1607-18.
[http://dx.doi.org/10.2741/1644] [PMID: 15769650]

[185] Balachandran P, Govindarajan R. Canceran ayurvedic perspective. Pharmacol Res 2005; 51(1): 19-30.
[http://dx.doi.org/10.1016/j.phrs.2004.04.010] [PMID: 15519531]

[186] Arnold JT, Wilkinson BP, Korytynski EA, Steel VE. Chemopreventive activity of Maharshi Amrit Kalash and related agents in rat tracheal epithelial and human tumor cells. Proc Am Assoc Cancer Res 1991; 32: 128-31.

[187] Sparano JA, Zhao F, Martino S, *et al.* Long-term follow-up of the E1199 Phase III trial evaluating the role of Taxane and schedule in operable breast cancer 2015.
[http://dx.doi.org/10.1200/JCO.2015.60.9271]

[188] Berlin J, King AC, Tutsch K, *et al.* A phase II study of vinblastine in combination with acrivastine in patients with advanced renal cell carcinoma. Invest New Drugs 1994; 12(2): 137-41.
[http://dx.doi.org/10.1007/BF00874444] [PMID: 7860231]

[189] Bertsch LA, Donaldson G. Quality of life analyses from vinorelbine (Navelbine) clinical trials of women with metastatic breast cancer. Semin Oncol 1995; 22(2) (Suppl. 5): 45-53.
[PMID: 7740333]

[190] Shinkai T, Eguchi K, Sasaki Y, *et al.* A randomised clinical trial of vindesine plus cisplatin *versus* mitomycin plus vindesine and cisplatin in advanced non-small cell lung cancer. Eur J Cancer 1991; 27(5): 571-5.
[http://dx.doi.org/10.1016/0277-5379(91)90220-8] [PMID: 1647183]

[191] Mross K, Richly H, Schleucher N, *et al.* A phase I clinical and pharmacokinetic study of the camptothecin glycoconjugate, BAY 383441, as a daily infusion in patients with advanced solid tumors. Ann Oncol 2004; 15(8): 1284-94.
[http://dx.doi.org/10.1093/annonc/mdh313] [PMID: 15277271]

[192] Falkson G, van Dyk JJ, van Eden EB, van der Merwe AM, van den Bergh JA, Falkson HC. A clinical trial of the oral form of 4-demethyl-epipodophyllotoxin-beta-D ethylidene glucoside (NSC 141540) VP 16213. Cancer 1975; 35(4): 1141-4.
[http://dx.doi.org/10.1002/1097-0142(197504)35:4<1141::AID-CNCR2820350418>3.0.CO;2-3] [PMID: 163675]

[193] Sharma RA, Euden SA, Platton SL, *et al.* Phase I clinical trial of oral curcumin: biomarkers of systemic activity and compliance. Clin Cancer Res 2004; 10(20): 6847-54.
[http://dx.doi.org/10.1158/1078-0432.CCR-04-0744] [PMID: 15501961]

CHAPTER 9

Translation of Natural Products into Clinically Effective Drugs: How Far We Have Gone

Ammad Ahmad Farooqi[1,*], Ilhan Yaylim[2], Rukset Attar[3], Muhammad Zahid Qureshi[4], Faiza Yasmeen[5] and Sobia Tabassum[6]

[1] *Institute of Biomedical and Genetic Engineering (IBGE), Islamabad, Pakistan*

[2] *Istanbul University Department of Molecular Medicine, Institute of Experimental Medicine Istanbul, Turkey*

[3] *Yeditepe University Medical School, Istanbul, Turkey*

[4] *GCU Department of Chemistry, Lahore, Pakistan*

[5] *Institute of Blood Transfusion Services, Lahore, Punjab, Pakistan*

[6] *Department of Bioinformatics and Biotechnology, International Islamic University, Islamabad, Pakistan*

Abstract: Data obtained from Human Genome Project has helped in transition of human diseases from a segmented view to a conceptual continuum. In accordance with this approach, identification of new gene targets has reinvigorated the field of natural product research and predominantly scientists are working to obtain these drugs through the use of high-throughput screening technologies and combinatorial chemistry. It is noteworthy that natural product templates combined with chemistry to selectively produce analogues will have higher chances of success. In this chapter we have attempted to summarize most recent advancements in clinical trials of natural products in different cancers. Keeping in view that structural variants contribute to the genomic landscape, multi-region whole-genome sequencing of hundreds of tumors will be helpful for a better understanding intra and inter-population genetic variability. Moreover, rapidly evolving field of nutrigenomics will play its part by tailoring the food or nutrition to the individual genotype. As we have developed deeper knowledge related to how wide ranging natural products modify cellular mechanisms, we may find that the continuum from pharmaceuticals to nutraceuticals through food-based biologically active phytochemicals will bring the disciplines of nutrigenomics and pharmacogenomics closer together.

Keywords: Apoptosis, Cancer, Metastasis, Phytochemicals, Signaling.

* **Corresponding author Ammad Ahmad Farooqi:** Institute of Biomedical and Genetic Engineering (IBGE), Islamabad, Pakistan; Tel/Fax: +92-334-4346213; E-mail: ammadfarooqi@rlmclahore.com

Sahdeo Prasad & Amit Kumar Tyagi (Eds.)

INTRODUCTION

Cancer is a complicated and genomically complex disease. Data obtained through high-throughput technologies has provided an ever-expanding list of regulators reported to be involved in cancer development, migration and invasion. Overexpression of oncogenes, inactivation of tumor suppressor genes, activation of pro-survival signaling cascades and loss of apoptotic cell death is some of the most extensively studied molecular mechanisms. Large-scale, cancer cell line-based screening of drug sensitivity has emerged as an important dimension of drug discovery and has proved to be helpful in complementing lower throughput, but complicated screens involving 3D and mixed tumors and stromal cultures, animal models and multi-targeted approaches. Off target effects and rapidly development of resistance against chemotherapeutic drugs is a major concern and explosion in genetic, genomic and proteomic information has opened new horizons to unfold the mystery of inter-individual differences in the body's ability to metabolize and response to nutrients.

PROSTATE CANCER

It is becoming progressively more understandable that genomic heterogeneity within individual prostate glands and between patients stems particularly from copy-number aberrations and structural variants. Subtypes of prostate cancers are being deeply explored using next-generation sequencing, but these subtypes are yet to be thoroughly investigated in the clinical setting for targeted screening and treatment. Research over the decades has added considerable information into the prostate cancer (PCa) biology. Dysregulation of intracellular signaling cascades and Prostate cancer stem cells have also been observed to play a key role in cancer progression and resistance against wide ranging therapeutics.

Dual PI3K/mTOR inhibitor NVP-BEZ235 effectively reduce the population of $CD133^+/CD44^+$ PCa progenitor cells [1]. NVP-BEZ235 and chemotherapeutic drug Taxotere effectively inhibited tumor growth in mice xenografted with prostate cancer cells [1]. $CD133^+CD44^{high}/AR^{-/low}$ side population (SP) cells isolated from tumorigenic and invasive WPE1-NB26 cells were noted to be resistant to docetaxel. Contrarily, docetaxel was effective against $CD133^{(-)}/CD44^{(low)}/AR^{(+)}$ non-side population cells isolated from the WPE1-NB26 cell line [2].

PC3 and DU145 cells combinatorially treated with NVP-BEZ235 and chemotherapeutic drug (Taxotere) had a 2-fold or greater decrease in $CD133^+/CD44^+$ progenitor cell populations [1]. NVP-BEZ235 in combination with either 5-FU or

Oxaliplatin 2-fold or greater decrease in the CD133⁻/CD44⁻ cell population [1]. NVP-BEZ235 and Taxotere synergistically induced 1.5-2.0-fold decrease in CD133⁻/CD44⁻ population [1]. Combining NVP-BEZ235, which preferentially targets progenitor populations, with chemotherapeutic drugs that target bulky tumors is more useful as compared to monotherapy [1]. Atorvastatin, a 3-hydroxy-3-methyl-glutarylcoenzyme-CoA (HMG-CoA) reductase inhibitor significantly inhibited α1, β1 integrins and phosphorylated levels of FAK and MYPT1. ROCK1 and FAK induced downstream signaling mediated diff-erentiation of CD133⁺CD44⁺ population derived from prostate cancer tissues however cellular differentiation was markedly inhibited upon atorvastatin treatment [3].

There is a recent evidence of efficacy evaluation of combinatorial therapy consisting of *Phellodendron amurense* bark extract (Nexrutine®) and rad-iotherapy in prostate cancer patients. The results revealed that treated patients did not show grade 3 toxicity, moreover, toxicities were detected transiently. Post-treatment data analysis indicated that 81% of the patients neoadjuvantly treated had a decline in PSA [4]. Statistically significant double-blind RCT has shown a noteworthy short-term effect on PSA in prostate cancer patients orally adm-inistered with a capsule consisting of a mixture of pomegranate, broccoli, turmeric and green tea [5].

Prostate tissue of cancer patients orally administered with pomegranate extract (POMx) was analyzed to study if systemically absorbed pomegranate extracts were converted into Urolithin A. Data indicated that significantly higher levels of Urolithin A levels were detected in POMx treated group. Moreover, 8-hydroxy-2'-deoxyguanosine (8-OHdG), an oxidative stress biomarker was also con-siderably reduced in POMx treated group [6]. Gene analysis of the specimen obtained from prostate cancer patient administered with short-term soy isoflavone revealed markedly down regulated genes including apoptotic protease activating factor-1 (APAF1), cell division cycle 27 (CDC27), cyclin B2 (CCNB2), cyclin C (CCNC), Ubiquitin-Activating Enzyme (UBE1), cyclin G1 (CCNG1), cyclin G2 (CCNG2), cullin 2 and cullin 3 [7].

Significantly higher levels of sulforaphane are present as compared to its inactive precursor, glucoraphanin. Both glucoraphanin and sulforaphane are known quantitatively in the administered broccoli sprouts. Moreover, glucoraphanin acts as a depot for the bio-active constituent so sulforaphane is slowly cleared from body. Since 2012 a pilot study is evaluating how broccoli sprout extract dose-dependently exerts biological effects on Dysplastic Nevi, as precursory lesions and malignant melanoma (clinicaltials.gov identifier: NCT01568996). Two other studies are examining efficacy in transitional cell bladder cancer patients preoperatively administered with broccoli sprout extract (BSE) (NLM identifier: NCT01108003) and intake of BSE in recurrent prostate cancer patients (clinicaltrials.gov identifier: NCT01228084).

Recently, a phase I clinical trial of muscadine grape skin extract in men with biochemically recurrent PCa was conducted. Findings revealed that 6 patients came off the study because of progression of disease (5 metastatic, 1 increase in levels of PSA) after a median exposure of 15 months however, 7 patients remained on study [8]. Dose-limiting toxicities were not observed and resultantly, 4,000 mg/d was considered as the highest dosage. Median within-patient "PSA Doubling Time (PSADT)" increased by 5.3 months. None of the patients experienced any sustained decline in the levels of PSA from baseline [8].

COLORECTAL CANCER

MB-6 is herbal preparation consisting of spirulina, *Antrodia camphorata* mycelia, curcumin extract and grape seed extract, green tea extract and fermented soybean extract. Seventy-two metastatic colorectal cancer patients were randomized for 16 weeks to receive 5-fluorouracil leucovorin and oxaliplatin combinatorially either with placebo or MB-6. Both disease progression rate and incidence of adverse events were considerably reduced in MB-6 treated group [9]. Colorectal cancer patients administered with pomegranate extract revealed 23 metabolites, but no ellagitannins, in urine, plasma, normal or malignant colon tissues. Levels of total and individual metabolites were higher in normal tissues as compared to malignant tissues [5]. Data obtained through a phase I study in colorectal cancer patients revealed that a Chinese herbal medicine PHY906 in combination with 5-fluorouracil, irinotecan and leucovorin induced a partial response in 4 patients and 7 had stable disease [10].

LUNG CANCER

Feitai Capsule, a Chinese herbal medicine has been used in combination with chemotherapy in NSCLC patients with stage IIIB/IV. Completion rate of chemotherapy was 96.42% in the treatment group, while it was 74.07% in control group [11]. Astragalus polysaccharide (APS), a polysaccharide isolated from the radix of *astragalus membranaceus* worked with notable efficacy in combination with cisplatin and vinorelbine in advanced NSCLC patients. Significant differences were noted in the overall patient quality of life after 3 cycles of treatment [12]. Aidi Injection®, a Chinese herbal preparation consisting of

Mylabri, *Radix astragali, Radix Acantropanacis* senticosi and *Radix ginsheng* is reported to be effective against stage III A NSCLC before surgical procedure. 100 mL of ADI was given to the patients by addition into 500 mL of 5% glucose injection which was intravenously dripped. There was a significantly higher efficacy rate in the treated patients as compared to control group [13].

HEPATOCELLULAR CARCINOMA

Jiedu granules, a Chinese herbal medicine compound has been shown to work effectively in combination with cinobufacini injection [14]. It has been observed that combinatorial treatment markedly delayed tumor recurrence and metastasis. Furthermore, both survival time and survival rate of postsurgical patients with HCC were enhanced [14].

Silybin phosphatidylcholine was tested for efficacy in 3 patients with advanced hepatocellular carcinoma (HCC) and serum concentrations of silibinin glucuronide and silibinin increased within 1 to 3 weeks. In all 3 patients, abnormalities of liver function and α-fetoprotein (tumor marker) increased, but after 56th day, a patient showed improvement in inflammatory biomarkers and abnormal liver function. Investigators were unable to calculate maximally tolerated dosage because patients died soon after enrollment [15].

PANCREATIC CANCER

Kanglaite Injection is a novel therapeutic agent reported to be produced from traditional Chinese medicinal herbs (the Coix Seed). It has been noted that gemcitabine in combination with Kanglaite Injection and radiotherapy improved median survival time in locally advanced pancreatic carcinoma patients [16]. Furthermore, CEA (ng/mL) and CA19-9 (U/mL) were respectively reduced to 7.41 +/- 2.37 and 118. 00 +/- 78.89 respectively in the treated patients [16].

Chinese botanical formula, PHY906, synergistically administered with capecitabine in pancreatic cancer patients resulted in median OS of 21.6 weeks and median PFS of 10.1 weeks [17]. Guben Yiliu II (GY II) combined with arterially perfused chemotherapeutic agent was noted to be clinically beneficial in patients of advanced pancreatic cancer. There was a notable amelioration in blood hypercoagulation state, elevated levels of cellular immunity and improved quality of life was noted in treated patients [18].

NATURAL PRODUCT INDUCED MODULATION OF EPIGENETIC MACHINERY

Epigenetic modifying ability of natural products has also been investigated in cell culture based and preclinical studies. Dujieqing Oral Liquid (DJQ) has been shown to reduce promoter methylation of O6-methylguanine-DNA methyl-transferase (MGMT) gene in middle-and-late stage tumor patients treated with chemotherapeutic drugs. Moreover, toxic and/or adverse reactions were milder in the DJQ treated group [19].

Ongoing Trials

Multivariate analysis verified that use of traditional Chinese medicine (TCM) remarkably reduced risk of all-cause mortality. Ban Zhi Lian, Bai Hua She She Cao and Huang Qi were noted to be most effective and frequently used TCMs [20]. Partial remission (PR) rate and Complete remission (CR) rate were significantly higher in esophageal cancer patients treated with oral *Fructus bruceae* oil (20 mL, 3 times per day for 12 weeks) and radiotherapy [21].

Clinical benefit rate in the cancer patients who received large dose compound *Sophora flavescens* Ait (20 ml/d) injection was 83% [22]. Scutellaria extract (SBE) enriched in baicalin treatment induced an increase in production of IFNγ in peripheral blood leukocytes, and reduced IL-10 and TNFα production in bone marrow cells (BMC), in acute lymphocytic leukemia patients [23].

80 patients with solid tumors who were on standard chemotherapeutic combinations were assigned randomly to bioavailability-enhanced curcuminoids preparations (180 mg/day). Curcuminoid supplementation correlated considerably with a significantly improved quality of life compared with placebo. Consistently, significantly reduced levels of inflammatory biomarkers and mediators (IL-6, MCP-1, CGRP, TNF-α, hs-CRP and substance P) had been observed. These clinical findings revealed that curcuminoids suppressed systemic inflammation [24].

Fuzheng Kang'ai decoction (FZKA), a 12-herb Chinese formula worked with effective synergy when combined with gefitinib to treat patients with patients with advanced non-small cell lung cancer (NSCLC). It was noted that combinatorially treated patients had prolonged progression-free survival (PFS) and median survival time (MST). Moreover, PFS in combinatorially treated patients was 13.1 months compared to 11.43 months in gefitinib treated patients [25].

Similarly, thoracoscopically administered dendritic cells and ginseng polysaccharides induced an increase in expression of the ratio of Th1/Th2 cytokines and Th1 cytokines [26]. Javanica oil emulsion injection (Yadanzi®) in combination with chemotherapeutic drugs worked with effective synergy and overall response rate of the synergistically treated group was high as compared to control group. Similarly, improved life rate was 82.8% in combinatorially treated group [27].

Mistletoe, *Viscum album* L., grows on different plants and particularly on deciduous trees, has been widely used both in traditional and complementary medicine. There was a significant regression of colon adenoma in a stage IIIC colon cancer patient intratumourally injected with *Viscum album* L extract [28]. Intratumoral injection of high-dose of VAE induced a notable decrease in tumor size in adenoid cystic carcinoma patient over a 10-month period [29]. Osteosarcoma patients were evaluated for Post relapse Disease-Free Survival (PRDFS) with Mistletoe or adjuvant oral etoposide. Results revealed that median PRDSF was 39 months in the Viscum group and 4 months in the Etoposide treated group [30]. Mistletoe has been tested for an improved efficacy by combining with different drugs. Combination treatment with Gemcitabine has been studied for efficacy in patients with advanced solid cancers. Drug tolerability and treatment compliance was higher in patients treated combinatorially with mistletoe and gemcitabine. MTD for gemcitabine was 1380 mg/m2 weekly on day one and eight of a 3-week cycle when co-administered with mistletoe 250 mg daily [31].

Locally advanced or metastatic pancreatic cancer patients responded differentially upon subcutaneous injections of *Viscum album* L. extract as evidenced by median Overall Survival (OS) of 3.4 *versus* 2.0 months within the 'poor' prognosis subgroup and 6.6 *versus* 3.2 months within subgroup of 'good' prognosis [32]. Patient of cutaneous squamous cell carcinoma responded to a high-dosage of peri-lesional *Viscum album* L. extract injections [33]. Pharmacologically it has been shown that frequency of adverse drug reactions (ADRs) associated with intratumorally injected mistletoe was 3 times and 5 times higher as compared to either intravenously or subcutaneously administered mistletoe. However, most frequently noted ADRs were associated with changes in body temperature and immune system. Intensity of ADRs ranged from mild (83.8%) to moderate (14.9%) [34].

CONCLUSION

Rapid advancements in field of analytical and phytochemistry, specific chemicals or a group of similar bioactive constituents from natural sources are being isolated/ extracted and investigated for identification of pharmacologically effective molecules [35]. However, it is relevant to mention that bioavailability associated issues of active principles of plants are major stumbling blocks because of poor oral bioavailability of chemicals which contained polyphenolic rings such as flavonoids and other H_2O soluble chemicals particularly, tanins and terpenoids. Poor bioavailability of these substances is mainly because of very low aqueous or lipid solubility, poor plasma membrane permeability and high molecular weight/size. Strategy of complexing water soluble constituents of plant extracts with phospholipids for improvement of their bioavailability was developed and patented as 'PHYTOSOME®' by an Italian pharmaceutical company. Phytosome® contained combinations of soy lecithin with polyphenolic compounds, which considerably enhanced their bioavailability and absorption.

For efficient translation of laboratory findings to clinically effective drugs, there should be careful screening of natural product extracts against purified enzymes and, in molecularly defined assays. Exclusion of 'distracting' molecules, such saponins and tannins, may significantly enhance chances of identification of compounds with genuine anticancer activity. In addition, designing of numerous analogues is helpful to extensively investigate structure–activity relationships, and better understanding of the signaling pathways targeted by these phytochemicals may prove to be useful before the choice of suitable pre-clinical candidates can be made.

Clinical feedback should be obtained quickly and carefully. This is achievable through use of relevant and more quantitative end points, and by utilizing imaging and molecular biomarker/s. Carefully designed clinical trials will be helpful to tactfully analyze molecular biology and genetics of the targets and pharmacology of the drugs. Moreover, it should be indicative of expected outcomes, which is often disease stabilization instead of rapid regression. Use of Biomarkers will be helpful in patient selection most likely to respond to different phytochemicals, to monitor therapy, to show proof-of-concept and optimally designed schedules. Use of pharmacologic audit trail may assist in decision making and risk management. Following section deals with most recent advancements in clinical trials of natural products in different cancers.

CONFLICT OF INTEREST

The authors confirm that they have no conflict of interest to declare for this publication.

ACKNOWLEDGEMENTS

The authors would like to pay their sincere thanks to Maira Mariam for English language editing.

REFERENCES

[1]   Dubrovska A, Elliott J, Salamone RJ, *et al.* Combination therapy targeting both tumor-initiating and differentiated cell populations in prostate carcinoma. Clin Cancer Res 2010; 16(23): 5692-702.
[http://dx.doi.org/10.1158/1078-0432.CCR-10-1601] [PMID: 21138868]

[2]   Mimeault M, Johansson SL, Henichart JP, Depreux P, Batra SK. Cytotoxic effects induced by docetaxel, gefitinib, and cyclopamine on side population and nonside population cell fractions from human invasive prostate cancer cells. Mol Cancer Ther 2010; 9(3): 617-30.
[http://dx.doi.org/10.1158/1535-7163.MCT-09-1013] [PMID: 20179163]

[3]   Rentala S, Chintala R, Guda M, Chintala M, Komarraju AL, Mangamoori LN. Atorvastatin inhibited Rho-associated kinase 1 (ROCK1) and focal adhesion kinase (FAK) mediated adhesion and differentiation of CD133+CD44+ prostate cancer stem cells. Biochem Biophys Res Commun 2013; 441(3): 586-92.
[http://dx.doi.org/10.1016/j.bbrc.2013.10.112] [PMID: 24177008]

[4]   Swanson GP, Jones WE III, Ha CS, Jenkins CA, Kumar AP, Basler J. Tolerance of Phellodendron amurense bark extract (Nexrutine®) in patients with human prostate cancer. Phytother Res 2015; 29(1): 40-2.
[http://dx.doi.org/10.1002/ptr.5221] [PMID: 25205619]

[5]   Nuñez-Sánchez MA, García-Villalba R, Monedero-Saiz T, *et al.* Targeted metabolic profiling of pomegranate polyphenols and urolithins in plasma, urine and colon tissues from colorectal cancer patients. Mol Nutr Food Res 2014; 58(6): 1199-211.
[http://dx.doi.org/10.1002/mnfr.201300931] [PMID: 24532260]

[6]   Freedland SJ, Carducci M, Kroeger N, *et al.* A double-blind, randomized, neoadjuvant study of the tissue effects of POMx pills in men with prostate cancer before radical prostatectomy. Cancer Prev Res (Phila) 2013; 6(10): 1120-7.
[http://dx.doi.org/10.1158/1940-6207.CAPR-12-0423] [PMID: 23985577]

[7]   Hamilton-Reeves JM, Banerjee S, Banerjee SK, *et al.* Short-term soy isoflavone intervention in patients with localized prostate cancer: a randomized, double-blind, placebo-controlled trial. PLoS One 2013; 8(7): e68331.
[http://dx.doi.org/10.1371/journal.pone.0068331] [PMID: 23874588]

[8]   Paller CJ, Rudek MA, Zhou XC, *et al.* A phase I study of muscadine grape skin extract in men with biochemically recurrent prostate cancer: Safety, tolerability, and dose determination. Prostate 2015; 75(14): 1518-25.
[http://dx.doi.org/10.1002/pros.23024] [PMID: 26012728]

[9]   Chen WT, Yang TS, Chen HC, *et al.* Effectiveness of a novel herbal agent MB-6 as a potential adjunct to 5-fluoracil-based chemotherapy in colorectal cancer. Nutr Res 2014; 34(7): 585-94.
[http://dx.doi.org/10.1016/j.nutres.2014.06.010] [PMID: 25150117]

[10]   Kummar S, Copur MS, Rose M, *et al.* A phase I study of the chinese herbal medicine PHY906 as a modulator of irinotecan-based chemotherapy in patients with advanced colorectal cancer. Clin Colorectal Cancer 2011; 10(2): 85-96.
[http://dx.doi.org/10.1016/j.clcc.2011.03.003] [PMID: 21859559]

[11]   Deng SQ, Ouyang XN, Yu ZY, *et al.* [Influence of Chinese herbal medicine Feitai Capsule on completion or delay of chemotherapy in patients with stage IIIB/IV non-small-cell lung cancer: a randomized controlled trial]. Zhong Xi Yi Jie He Xue Bao 2012; 10(6): 635-40.
[http://dx.doi.org/10.3736/jcim20120606] [PMID: 22704411]

[12]   Guo L, Bai SP, Zhao L, Wang XH. Astragalus polysaccharide injection integrated with vinorelbine and cisplatin for patients with advanced non-small cell lung cancer: effects on quality of life and survival. Med Oncol 2012; 29(3): 1656-62.
[http://dx.doi.org/10.1007/s12032-011-0068-9] [PMID: 21928106]

[13]   Sun XF, Pei YT, Yin QW, *et al.* Application of Aidi injection in the bronchial artery infused neo-adjuvant chemotherapy for stage III A non-small cell lung cancer before surgical operation. Chin J Integr Med 2010; 16: 537-41.
[http://dx.doi.org/10.1007/s11655-010-0569-y] [PMID: 21110180]

[14]   Chen Z, Chen HY, Lang QB, *et al.* Preventive effects of jiedu granules combined with cinobufacini injection *versus* transcatheter arterial chemoembolization in post-surgical patients with hepatocellular carcinoma: a case-control trial. Chin J Integr Med 2012; 18(5): 339-44.
[http://dx.doi.org/10.1007/s11655-012-1083-1] [PMID: 22549390]

[15]   Siegel AB, Narayan R, Rodriguez R, *et al.* A phase I dose-finding study of silybin phosphatidylcholine (milk thistle) in patients with advanced hepatocellular carcinoma. Integr Cancer Ther 2014; 13(1): 46-53.
[http://dx.doi.org/10.1177/1534735413490798] [PMID: 23757319]

[16]   Shen WS, Shu ZQ, Deng LC. [Curative effect of 3D-CRT combined with gemcitabine concurrently with addition of Kanglaite Injection in treatment of locally advanced pancreatic]. Zhongguo Zhong Xi Yi Jie He Za Zhi 2012; 32(7): 902-5.
[PMID: 23019943]

[17]   Saif MW, Li J, Lamb L, *et al.* First-in-human phase II trial of the botanical formulation PHY906 with capecitabine as second-line therapy in patients with advanced pancreatic cancer. Cancer Chemother Pharmacol 2014; 73(2): 373-80.
[http://dx.doi.org/10.1007/s00280-013-2359-7] [PMID: 24297682]

[18]   Zhang Q, Wang XM, Chi HC. [Effects of Guben Yiliu II combined with arterial perfusion with chemotherapeutic agent in treating advanced pancreatic cancer]. Zhongguo Zhong Xi Yi Jie He Za Zhi 2007; 27(5): 400-3.
[PMID: 17650790]

[19]   Rong Z, Xu Y, Mo CM. [Effects of dujieqing oral liquid on the promoter methylation of the MGMT gene in middle-and-late stage tumor patients receiving chemotherapy]. Zhongguo Zhong Xi Yi Jie He Za Zhi 2012; 32(12): 1611-5.
[PMID: 23469597]

[20]   Lee YW, Chen TL, Shih YR, *et al.* Adjunctive traditional Chinese medicine therapy improves survival in patients with advanced breast cancer: a population-based study. Cancer 2014; 120(9): 1338-44.
[http://dx.doi.org/10.1002/cncr.28579] [PMID: 24496917]

[21]   Shan GY, Zhang S, Li GW, Chen YS, Liu XA, Wang JK. Clinical evaluation of oral Fructus bruceae oil combined with radiotherapy for the treatment of esophageal cancer. Chin J Integr Med 2011; 17(12): 933-6.
[http://dx.doi.org/10.1007/s11655-011-0953-2] [PMID: 22139545]

[22] Li DR, Lin HS. [Safety and effectiveness of large dose compound Sophora flavescens Ait injection in the treatment of advanced malignant tumors]. Zhonghua Zhong Liu Za Zhi 2011; 33(4): 291-4.
[PMID: 21575502]

[23] Orzechowska B, Chaber R, Wiśniewska A, *et al.* Baicalin from the extract of Scutellaria baicalensis affects the innate immunity and apoptosis in leukocytes of children with acute lymphocytic leukemia. Int Immunopharmacol 2014; 23(2): 558-67.
[http://dx.doi.org/10.1016/j.intimp.2014.10.005] [PMID: 25448499]

[24] Panahi Y, Saadat A, Beiraghdar F, Sahebkar A. Adjuvant therapy with bioavailability-boosted curcuminoids suppresses systemic inflammation and improves quality of life in patients with solid tumors: a randomized double-blind placebo-controlled trial. Phytother Res 2014; 28(10): 1461-7.
[http://dx.doi.org/10.1002/ptr.5149] [PMID: 24648302]

[25] Yang XB, Wu WY, Long SQ, *et al.* Fuzheng Kangai decoction combined with gefitinib in advanced non-small cell lung cancer patients with epidermal growth factor receptor mutations: study protocol for a randomized controlled trial. Trials 2015; 16: 146.
[http://dx.doi.org/10.1186/s13063-015-0685-2] [PMID: 25873045]

[26] Ma J, Liu H, Wang X. Effect of ginseng polysaccharides and dendritic cells on the balance of Th1/Th2 T helper cells in patients with non-small cell lung cancer. J Tradit Chin Med 2014; 34(6): 641-5.
[http://dx.doi.org/10.1016/S0254-6272(15)30076-5] [PMID: 25618966]

[27] Lu YY, Huang XE, Cao J, *et al.* Phase II study on Javanica oil emulsion injection (Yadanzi®) combined with chemotherapy in treating patients with advanced lung adenocarcinoma. Asian Pac J Cancer Prev 2013; 14(8): 4791-4.
[http://dx.doi.org/10.7314/APJCP.2013.14.8.4791] [PMID: 24083745]

[28] von Schoen-Angerer T, Goyert A, Vagedes J, Kiene H, Merckens H, Kienle GS. Disappearance of an advanced adenomatous colon polyp after intratumoural injection with Viscum album (European mistletoe) extract: a case report. J Gastrointestin Liver Dis 2014; 23(4): 449-52.
[PMID: 25532007]

[29] Werthmann PG, Helling D, Heusser P, Kienle GS. Tumour response following high-dose intratumoural application of Viscum album on a patient with adenoid cystic carcinoma. BMJ Case Rep 2014.
[http://dx.doi.org/10.1136/bcr-2013-203180]

[30] Longhi A, Reif M, Mariani E, Ferrari S. A Randomized Study on Postrelapse Disease-Free Survival with Adjuvant Mistletoe *versus* Oral Etoposide in Osteosarcoma Patients. Evid Based Complement Alternat Med 2014.
[http://dx.doi.org/10.1155/2014/210198] [PMID: 210198]

[31] Mansky PJ, Wallerstedt DB, Sannes TS, *et al.* NCCAM/NCI Phase 1 Study of Mistletoe Extract and Gemcitabine in Patients with Advanced Solid Tumors. Evid Based Complement Alternat Med 2013.
[http://dx.doi.org/10.1155/2013/964592] [PMID: 24285980]

[32] Tröger W, Galun D, Reif M, Schumann A, Stanković N, Milićević M. Viscum album [L.] extract therapy in patients with locally advanced or metastatic pancreatic cancer: a randomised clinical trial on overall survival. Eur J Cancer 2013; 49(18): 3788-97.
[http://dx.doi.org/10.1016/j.ejca.2013.06.043] [PMID: 23890767]

[33] Werthmann PG, Sträter G, Friesland H, Kienle GS. Durable response of cutaneous squamous cell carcinoma following high-dose peri-lesional injections of Viscum album extractsa case report. Phytomedicine 2013; 20(3-4): 324-7.
[http://dx.doi.org/10.1016/j.phymed.2012.11.001] [PMID: 23394841]

[34]   Steele ML, Axtner J, Happe A, Kröz M, Matthes H, Schad F. Use and safety of intratumoral application of European mistletoe (Viscum album L) preparations in Oncology. Integr Cancer Ther 2015; 14(2): 140-8.
[http://dx.doi.org/10.1177/1534735414563977] [PMID: 25552476]

[35]   Khan J, Alexander A, Ajazuddin , *et al.* Recent advances and future prospects of phyto-phospholipid complexation technique for improving pharmacokinetic profile of plant actives. J Control Rel 2013; 168(1): 50-60.
[http://dx.doi.org/10.1016/j.jconrel.2013.02.025] [PMID: 23474031]

SUBJECT INDEX

A

Abnormal cells 129, 147, 149
Acanthopanax Senticosus 106, 112
A-carotene 70
Acid(S) 6, 8, 13, 35, 36, 37, 39, 40, 41, 42, 43, 44, 45, 46, 47, 48, 50, 51, 52, 53, 75, 76, 93, 94, 95, 96, 111, 133, 160,
  11-keto-β-boswellic 35, 37, 41, 43, 44, 45, 46, 47, 50, 51
  α/β-boswellic 35, 37, 39
  α-boswellic 35, 36, 39
  β-boswellic 36, 37, 43, 44, 48, 50, 51, 53
  O-acetyl-11-keto-β-boswellic 37, 40, 41
  O-β-D-xylopyronosylellagic 96
  acetyl-11-keto-β-boswellic 37, 39, 51
  Acyl derivatives of boswellic 40, 42
  arachidonic 160
  ascorbic 111
  betulinic 6, 75, 76
  ellagic 75, 95
  fatty 9, 75, 133
  hydroxamic 52
  ursolic 6, 8, 13
Acquired immune deficiency syndrome (AIDS) 88
Actinomycins 165
Activities 21, 26, 32, 40, 43, 46, 50, 51, 87, 92, 94, 133, 153, 158
  anti-arthritic 32, 50, 51
  anti-proliferation 40, 46
  antiproliferative 92, 94
  biological 21, 26, 43, 87, 133, 153
  higher 158
Adenocarcinoma 72
Adverse drug reactions (ADRs) 188
Agelasine 154
Agents 65, 75, 76, 80, 81, 149, 150, 165, 166, 167
  anti cancer 167
  natural 75, 81
  new anticancer 150
AKBA in PC-3 cancer cell lines 48
Alkaloids 90, 91, 126, 133, 134, 147, 151, 153, 154

Allanxanthone 95
Alliinase 21, 22
Allyl disulfide 23
Allyl methyl sulfide (AMS) 23
Aloe vera mouthwash 115, 116
Amaranthaceae 107
Amaranthus paniculatus 107, 108
Angiogenesis 40, 76, 77, 130, 153, 161, 162, 163
Anthracycline 165
Anti-cancer 32, 35, 38, 43, 147, 160, 162, 166
Anti-cancer activities 5, 6, 13, 20, 21, 25, 41, 44, 48, 49, 52, 66, 71, 75, 89, 90, 91, 92, 95, 98, 112, 115, 116, 126, 130, 134, 138, 153, 162, 164, 167, 189
  exhibited potential 90
  illustrated 167
Anti-cancer agent 12, 71, 98, 115, 125, 152, 153, 165
  derived 152, 153, 165
  potential 98
Anti-cancer cell lines 90
Anti-cancer compounds 4, 131, 134, 136, 137, 165
  clinical 134
Anti-cancer drugs 32, 87, 134, 137, 138, 162
  new 32, 87
  plant-derived 162
  potential 87
Anti-cancer drugs development 133
Anti-cancer effects 10, 12, 105, 108, 134, 136, 163
  displayed 12
  exhibited 10
Anti-cancer marine compounds 131
Anti-cancer targets, validated 99
Anti-cancer therapy 153
Anti-inflammatory activities 5, 6, 20, 71, 135
Anti-inflammatory drugs 50
Antioxidant effect 24, 113
Anti-proliferative activity 39, 40, 41, 108, 134, 135, 136, 137, 138
Anti-tumor activity 38, 134, 137, 157, 159, 162
Apoptosis 6, 9, 12, 25, 26, 39, 40, 41, 42, 43, 45, 47, 49, 52, 66, 69, 70, 71, 72, 73, 76,

www.ingramcontent.com/pod-product-compliance
Lightning Source LLC
Chambersburg PA
CBHW041727210326
41598CB00008B/799